Mixing of Solids

Particle Technology Series

Series Editor

Professor Brian Scarlett
Technical University of Delft

The Kluwer Particle Technology Series of books is the successor to the Chapman and Hall Powder Technology Series. The aims are the same, the scope is wider. The particles involved may be solid or they may be droplets. The size range may be granular, powder or nano-scale. The accent may be on materials or on equipment, it may be practical or theoretical. Each book can add one brick to a fascinating and vital technology. Such a vast field cannot be covered by a carefully organised tome or encyclopaedia.Any author who has a view or experience to contribute is welcome. The subject of particle technology is set to enjoy its golden times at the start of the new millennium and I expect that the growth of this series of books will reflect that trend.

Mixing of Solids

RALF WEINEKÖTTER
HERMANN GERICKE

Gericke Ltd.,
Regensdorf/Zürich, Switzerland

KLUWER ACADEMIC PUBLISHERS
DORDRECHT / BOSTON / LONDON

A C.I.P. Catalogue record for this book is available from the Library of Congress.

ISBN 0-7923-6229-2

Published by Kluwer Academic Publishers,
P.O. Box 17, 3300 AA Dordrecht, The Netherlands.

Sold and distributed in North, Central and South America
by Kluwer Academic Publishers,
101 Philip Drive, Norwell, MA 02061, U.S.A.

In all other countries, sold and distributed
by Kluwer Academic Publishers,
P.O. Box 322, 3300 AH Dordrecht, The Netherlands.

Printed on acid-free paper

Printed in the Netherlands.

Contents

Preface

This book is a welcome edition to the Particle Technology Series, formerly Powder Technology Series. It is the second book in the series which describes powder mixing and we make no excuses for that. The topic of powder mixing is fundamental to powder technology and is one which always aroses interest. That will not change. As powder products become more complex they will pose new mixing problems. The solutions lie in the intelligent use of equipment, an understanding of powder properties and a good knowledge of basic statistics. The authors of this book have presented those three ingredients with great clarity. The book is based on long experience and deep thought, I have enjoyed reading it and am pleased to recommend it.

Delft University of Technology, NL-Delft, July 1999

Brian Scarlett, Series Editor

Foreword to the English Edition

In response to many enquiries from industrial organisations and institutes involved with the technology of processing bulk materials, we are pleased to present the English edition „Mixing of Solids". The text is based on the first edition written in the German language „Mischen von Schüttgütern" [1].
The English version has been revised and updated to include the most recent research developments, the conclusions of which have been taken into account.
We express our sincerest thanks to the editor of this edition Professor Brian Scarlett, of the UNIVERSITY OF DELFT (NL). We hope that the information contained in within this finds the same strong interest as the German edition.

CH-Regensdorf-Zurich, July 1999

Ralf Weinekötter Hermann Gericke

Foreword to the First Edition

The mixing of solids, powders, granules, fibres and flakes represent one of the essential basic processes for a diverse range of industries. The processing of foodstuffs, pharmaceuticals, chemicals, plastics and intermediate materials rely upon the efficient bringing together of the various elements of the recipe.
This book deals with the theoretical and practical aspects for the mixing of solids and gives an overview presentation of the subject.
For many years the mixing of solids together with milling and separation processes has been a key consideration for producers of processed goods. The development of equipment used for mixing can be traced back well over 100 years from the Gericke company documentation and archives.

At the end of the 19th century recirculating gravity mixers were used in conjunction with continuous bucket and chain elevators for large volume production and manually operated or slave drum mixers for small volumes. The vertical worm driven machines which followed these units relied upon the free flowing characteristics of the products.
It was not until the introduction of forced mixers in the form of horizontal ribbon blenders that other non free flowing cohesive powders could be successfully mixed. Mixing times were later shortened by the use of rotating paddle, plough share and air mixers.
Vertical mixers with rotating worm drives also blend large batches of product with poor flow characteristics. High efficiency twin shaft horizontal blenders further reduced mixing times and minimised the energy input.
Over the past decade the use of continuous mixing together with improvements in constant feeding technology has become more frequently used and has opened up more possibilities for the processor.
The collaboration of the two authors stems for a current research project in the field of solids mixing which under the leadership of Professor Lothar Reh, led to a further collaboration between the Institute for Mechanical Process Technology at the SWISS FEDERAL INSTITUTE OF TECHNOLOGY in Zurich, GERICKE LTD Regensdorf - Zurich and LONZA LTD of the ALUSUISSE LONZA GROUP. As a result of this research the conclusions, methods and guidelines of Professor Reh have been reflected in this book.

CH-Regensdorf-Zurich, February 1995

Ralf Weinekötter Hermann Gericke

Index of Symbols

Symbol	Description	Unit
a	Exponential autocorrelogram coefficient	Hz
A	Proportionality factor	-
AD	Feeder flow rate average deviation	kg/h
Bo	Bodenstein number	-
c_p	Tracer component variability	-
c_q	Variability of the other ingredients	-
c_w	Drag coefficient	-
C_{xx}	The autocovariance function of x	-
d	Mixer diameter	m
d_p	Particle diameter	m
$d_{p3,xx}$	xx% value of the cumulative distribution of the volumetric equivalent diameter	m
$d_{p,v}$	Ratio of the diameter of coarse and fine particles	-
$d_{p,p}$ $d_{p,q}$	Particle diameter of ingredients p and q	m
d_z	Multiple of the displacement related to cell length	
$\bar{d}$	Integer part of d_z	
D	Axial coefficient of dispersion	m/s^2
$E\{\}$	Anticipated value	-
$E(t)$	Residence time distribution density function	-
E_{Mix}	Mixing energy	W
f	Frequency	Hz
f	System error	-
$\hat{f}$	t_v/T_p	-
f_k	Discrete frequency ($=k\Delta f$)	Hz
$F(t)$	Cumulative distribution function of the residence time	
g	Gravitational acceleration	m^2/s
G_{xx}	Single-sided power density spectrum; Fourier transformed C_{xx}	s
$\hat{G}_{xx}$	Estimated value for G_{xx}	s
H	Height of the fluidized bed	m
I_H	Number of fine particles in the agglomerate	-
J_H	Number of coarse particles in the agglomerate	-
K	Coefficient of absorption	-

l_0	Linear scale of segregation	s (m)
l_0^m	normalised scale of segregation	kg
l_s	immersion length of the screw of a concial screw mixer	m
L	Mixer length	m
m	Index, mean	-
m*	Total feed	kg/s
$\dot{m}$, m°	Mass flow rate	kg/s
$\overline{\dot{m}}$	Mass flow rate average value	kg/s
m_f	Mass of a fine particle	kg
m_p, m_q	Average particle weight of the two components in the mixture	kg
M	Coefficient of mixing	m^2/s
M	Mass of a sample	kg
M	Mass of a batch	kg
M_b	Weight of the whole mixture	kg
n	Random sampling scope	-
n	Time level	-
n	Rotational frequency	Hz
n_A	Rotational frequency of the arm of a conical screw mixer	Hz
n_b	Refraction index	-
n_d	T_d/T	-
n_p	Number of particles in the sample	-
n_s	Rotational frequency of the screw of a conical screw mixer	Hz
N	Multiple of the base value	-
N	Rotational frequency	Hz
N.A.	Numerical aperture	-
Ne	Newton number	-
N_g	Number of samples in the basic whole	-
N_R	Number of ideal stirred vessels in the model	-
N_t	Number of agglomerates in the sample	-
p	Tracer component concentration in the basic whole	-
p_g	Proportional mass volume of the coarse ingredient	-
P	Power	W
q	1-p	-
Q_3^c	Grain size distribution of the coarser ingredient	-
r	Mixer radius	m
r^2	Coefficient of correlation	-
R	Relative signal	-
R_∞	Relative signal reflected by an infinitely dense layer of powder	-
$\hat{R}$	Residue	-
R_{xx}	Autocorrelation function of x	-
$R_{xx,feed}$	Autocorrelation function of *x* in the feed	-
s	Horizontal trajectory	m

S	Empirical standard deviation	-
S^2	Random sample variance	-
S_D	Total feeding error under long term observation	kg/h
S_{rel}	Relative feeding constant	-
t, t'	Time	s
t_V	mean residence time	s
t_f, t_m, t_e, t_i	Filling, mixing, discharging and idle time	s
$t_{w,n-1}$	Student faktor	-
t^*	Mixing Time	-
T	Measurement period	s
T_d	Multiple of T	s
T_p	Feed fluctuation period	s
v	Axial velocity	m/s
v_0	Horizontal velocity	m/s
V*	Volumetric total feed	m^3/s
VRR	Variance reduction ratio = $\sigma_{in}^2/\sigma_{out}^2$	-
VRR_{fluid}	Variance reduction ratio for gases or fluids	-
VRR_{solids}	Variance reduction ratio for solids	
W{}	Probability	-
w_f	Weight of a particle of the fine ingredient	kg
x	Concentration of the tracer component	-
x_0	Amplitude of the concentration's fluctuation	-
x_1	Initial concentration in a jump test	-
x_2	Final concentration in a jump test	-
$x_{i,n}$	Concentration in the 'i' element, time level n	
$\tilde{x}$	Average concentration in two neighbouring cells	-
$\hat{x}$	x-μ	-
X_k	Fourier transformation of concentration x	-
$\hat{X}_k$	" of $\hat{x}$	
z	Spatial coordinate	m
z	Multiple of the release velocity	-
z_n	Natural number	-

δ	Delta function	-
Δf	Discrete frequency interval: $\Delta f = 1/T$	Hz
Δt	Time interval between 2 measurements	s
ξ	z/L	-
η	Dynamic viscosity	Pas
Θ	t/tv	-
ϑ	Angle of refraction	-

$\vartheta(i,j)$	The probability that an agglomerate contains fine 'I' and coarse 'j' particles	-
μ	Random sample mean concentration value	-
ρ	Density of the solids	kg/m^3
ρ_{bulk}	Bulk density	kg/m^3
ρ_{xx}	Autocorrelogram of x	-
$\rho_{xx}(\tau)$	Autocorrelation coefficient of x	-
ρ_s	Density of the solids	kg/m^3
σ_p, σ_q	Standard deviation of the particle weight for the two ingredients in the mix	kg
σ^2	Variance	-
σ^2_z	Variance of a random mix	-
σ_1^2	Variance obtained from random samples of a fixed basic size	-
$\sigma^2_{end,solids}$	Variance of x at the outlet from a continuous solids mixer during time constant feeding	-
σ^2_{in} ; σ^2_{out}	Variance of the tracer concentration at the mixer's inlet and outlet	-
σ_N^2	Variance obtained from random samples 'N' times the basic size	-
σ^2_{tv}	Residence time variance	s^2
σ_E^2	Variance after the mixing process has ended – in an ideal case this is identical to an ideal random mix	-
σ_M^2	Variance caused by inaccuracy in the measurements	-
$\sigma^2_{syst}(t)$	System deviation from a mix status	-
τ	Temporal or local spatial interval between 2 samples or measurements	s or m
$\widetilde{\tau}$	τ/t_V	-
$\Phi(\chi^2)$	Cumulative function of	
χ^2	Chi square distribution	-
χ^2_l ; χ^2_u	Chi square distribution variables. In a two-sided confidence interval 'l' stands for the lower and 'u' for the upper limit	-
ω	Angular velocity	1/s

1 Introduction

All material conversion processes can be split into three categories:
- *Mixing*
- *Reactions, or the conversion of substances*
- *Separation*

Mixing unites separate substances into a blend. Similarly to the case of separation processes - those of mixing in reverse, extending from the mechanical sifting of solids to the thermal distillation of liquids - the abundance of applications is extraordinarily large. Reaction technology is an important field of application since the prerequisite of each reaction is an effective stoichiometric mixing of the reactive components. The premix burner and the stir boiler are examples showing the significance of mixing in reaction technology. With premix burners the mixing task is shown in the description of the reactor model. In the chemicals industry stir boilers serve both as reactors and to premix liquids and suspensions. Poor premixing of the components reduces conversion of the reaction's educts and has an adverse effect upon the process's efficiency. In addition to reaction technology mixing is a pivotal process operation in wide areas of the material conversion industry. Uhl and Gray's definition [2] shows the multi-faceted nature of mixing:

The term mixing is applied to operations which tend to reduce nonuniformities or gradients in composition, properties, or temperature of material in bulk. Such mixing is accomplished by movement of material between various parts of the whole mass.

By far the most important use of mixing is the production of a homogenous blend of several ingredients and thus to neutralize concentrations. But if the volume of material consists of one ingredient or compound exhibiting fluctuating properties caused by the production process, or on account of its raw material, the term homogenization is used for the neutralization of these fluctuations. However, this term is also applied to the mixing of soluble liquids.

The various mixing systems are normally categorized according to their principle phase (see Fig. 1.1). This extends from the single step mixing of fluids (fluid: gas or liquid), gases with liquids (dispersing and atomizing), solids with liquids/gases (fluidizing and suspending), to the pure mixing of solids with solids. Under solids mixing are also counted systems which have only a small proportion of liquids and where after the mixing process that liquid is absorbed into the solids (wetting and moistening). With increased proportions of liquid one talks about kneading. Very strong shear forces have to be applied if the viscosity is very high and the mix is shaped plastically. Here kneaders, internal mixers or continuously operated extruders are used. In 2-step systems inertial and shear forces enlarge the surfaces making contact. This shows that a primary application of mixing is not in itself the dispersal of solids or a gas in the fluid phase but also serves to improve heat and material exchange.

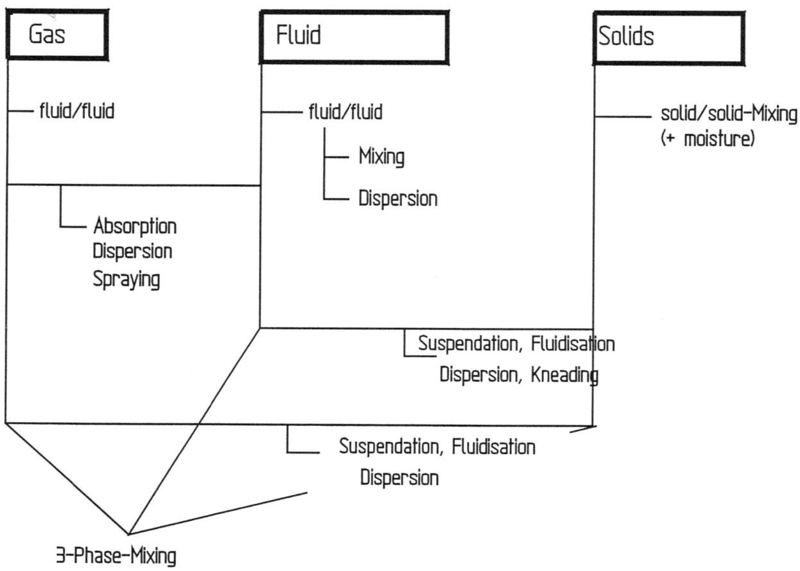

Fig. 1.1 The multi-faceted nature of mixing

The technical process of mixing is performed by a multitude of equipment available on the market. However, mixing processes are not always designed with the appropriate care. This causes a significant financial loss which arises in two ways:

a) The quality of the mix is poor: Where the mixing produces the end product this will be noticed immediately at the product's quality inspection. Frequently, however, mixing is only one in a whole series of further processing stages. Here the effects of unsatisfactory blending are less apparent.

b) The homogeneity is satisfactory but the effort employed is too great (OVERMIXING): Overmixing in batch blending is induced by an overlong mixing time or a too long residence time in the case of continuous blending. This leads to increased strain on the mixture which can have an adverse effect on the quality of sensitive products. Furthermore, larger or more numerous pieces of equipment must be used than would be necessary in the case of an optimally configured mixing process.

The smallest units which can be moved by mixing are molecules in the case of gases and liquids, whereas with solids it is their individual particles, the dimensions of which exceed the diameter of a molecule by orders of magnitude. Two different blending mechanisms are differentiated: *convection and diffusion.*

Convection is taken to mean the common movement of relatively large groups of elements as produced for example by agitators, pumps or rotating devices. Divisional processes are used because of speed gradients, i.e. convection progressively decreases the size and intervals between unmixed areas in the mixer.

Diffusion is taken to mean the individual, random movement of individual elements. In the case of fluid mixtures this is induced by Brown's molecular movement. Since motion of solid particles has to be induced by external energy input, „diffusion“ of the particles only occurs when the material mass is in motion.

Convection and diffusion occur simultaneously in most mixers. Large scale mix equalisation is effected by convection *(macromixing)* and fine blending *(micromixing)* by diffusion.

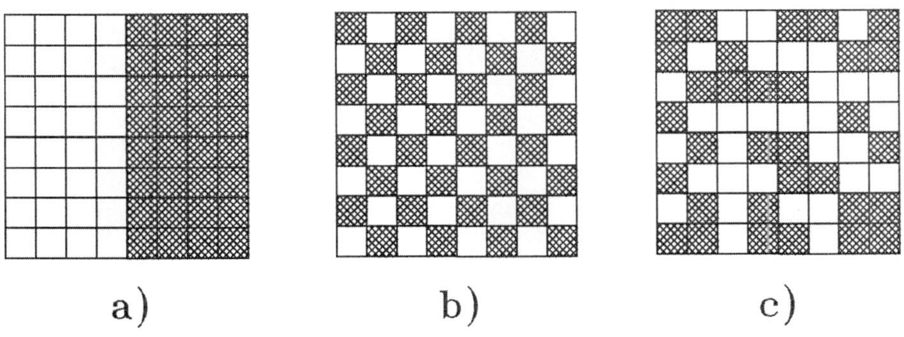

Fig. 1.2 Idealised mixing conditions: **a)** completely segregated **b)** ordered **c)** random

The product of a blending process is not a regular arrangement of the elements. Because of the random nature of the molecules' movement, conglobates of fluids or particles will at best only produce a random distribution of the individual particles amongst each other (see Fig. 1.2). This *random nature* shows up particularly clearly when blending solids, since the dimensions of the samples taken and of the elements (particles) do not greatly differ. Thus for example in solids mixing, even with an infinitely long mixing time, the concentration of a tracer element in a set of samples is randomly distributed around an average value whilst in fluid mixing this kind of sample usually contains many elements (molecules) so that the random nature is no longer detectable and consequently the sample concentration is identical to that of the average.

When analysing the blending process the following three questions must be answered:

- *How good is the mixing?*
- *How quickly will this mix status be reached?*
- *How high is the required energy input?*

Fig. 1.3 Mixing for the Extrusion of PVC-pipes - The photo shows totally eight gravimetric feeders which guarantee a controlled feed of two continuous mixers which supply their intermediate products to two downstream installed extruders.

The mixing of solids occupies a special place in the field of blending technology, since as of now these three questions can only be answered experimentally with regard to the *solids mixing process*. As will be shown again in the following chapters these experiments require considerable resources and call for statistical methods of analysis. This special position is grounded in the difficult to grasp parameters of materials. What is the value which designates a powder's mixability?

Industrial relevance of solids mixing

The mixing of powders, particles, flakes, granules gains increasingly economical importance in many different industries ranging from the mixing of human and animal foodstuff, pharmaceutical products, detergents, chemicals, plastics etc. In most cases the mixing process adds significantly value to the product, it can be regarded as a key process.

By mixing a new product or intermediate is created, the quality and such the price of this product is very often dependant on the efficiency of the mixing process. It will be shown that this efficiency is determined both by the materials to be mixed, e.g. particle size and particle size distribution, density, surface and the process and equipment used for performing the mixing. The design and operation of the mixing unit itself has a strong influence on the quality produced but upstream process steps like feeding, sifting, weighing, transport etc. determine also both quality and capacity of the mixing process. Downstream processing may destroy the product quality due to segregation (demixing). Continuous mixing is one solution which limits segregration by avoiding storage equipment.

Not all fields of application in respect to the blending of solids can be dealt with inside the framework of this book nor all the comprehensive scientific literature be presented. Rather more, general principles are pointed out and this book's target readership, students of process technology or chemical engineering, such as engineers and chemists, are given notes and sources of literature will help them in designing and checking solids mixing processes.

2 Mixing and Segregation Processes

As already prior mentioned solids have to be moved for achieving a blending. But as it will be shown that this forced movement may have the adverse effect - segregation. The „moveability“ is determined by parameters like particle size, density, surface roughness etc, there is a chance that with increased mixing time the indivual ingredients accumulate due to the individual movement of each component. This effect is called segregation or demixing. Many mixing processes are overlaid with a segregation process which obstructs the production of an ideal random mixture and conditions of equilibrium between mixing and segregation occur in the blenders. The segregation effect in mixers can be minimised by choosing an appropriate mixer type, but special care has to be taken for limiting segregation during the downstream handling of the mixture. The ingredients's selective motion may cause the mixture to separate out again during further processing or whilst being transported.

2.1 Dispersive and Convective Mixing

The mixing process can be observed in diagrammatic form as an overlapping of *dispersion* and *convection* (Fig. 2.1). Movement of the material is a prerequisite of both mechanisms.

Dispersion is understood to mean the completely random change of place of the individual particles. The frequency with which the particles of ingredient 'A' change places with those of another is related to the number of particles in the other ingredients in the direct vicinity of the particles in ingredient 'A'. Dispersion is therefore a local effect (micromixing) taking effect in the case of premix systems where a number of particles of different ingredients are in proximity and which leads to a fine mix in very small areas. If the ingredients are spatially separated at the beginning of the process it would take a long time to mix them through dispersion alone, since there is a very low number of different types of „neighbours“. Dispersion corresponds to diffusion in liquid mixtures. However in contrast to diffusion, mixing in the case of dispersion is not caused by any concentration gradient. An enforced movement produces the random dispersive mixing processes.

Convection causes a movement of larger groups of particles relative to each other (macromixing). The whole volume of material is continuously divided up and then mixed again after the portions have changed places (Fig. 2.1). The size of the groups, which are composed of just one ingredient and are consequently completely unmixed, is thus continuously reduced. Convection increases the number of assorted "neighbours" and thereby promotes the diffusive exchange processes. A material mass is divided up or convectively mixed through the rearrangement of a solid's layers by rotating devices in the mixer or by the fall of a stream of material in a static gravity mixer.

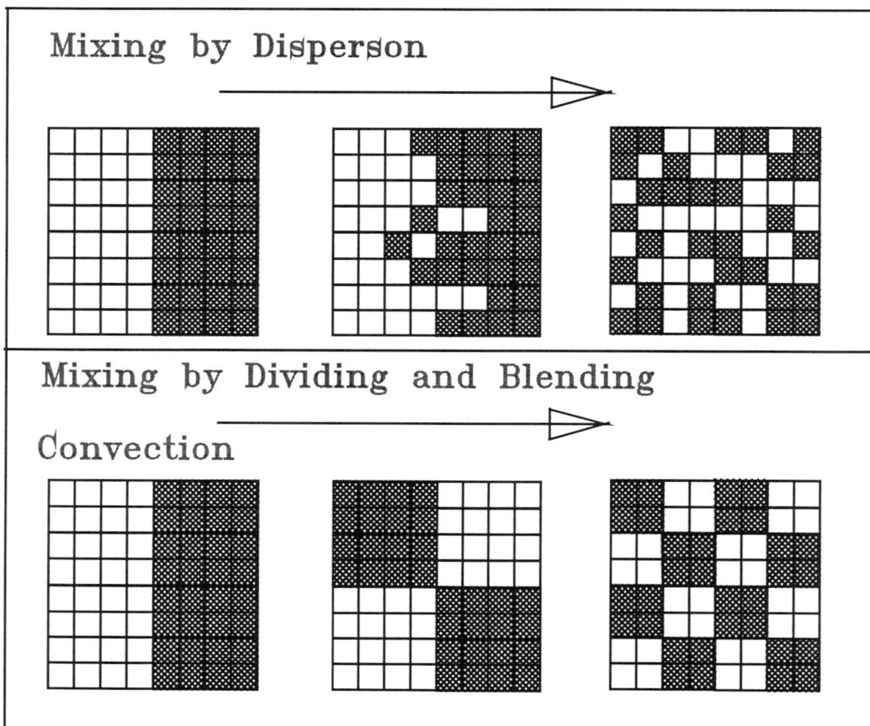

Fig. 2.1 The mechanism of mixing

2.2 Segregation in Solids

If the ingredients in a solids mixture possess a selective, individual motional behaviour the mixture's quality can be reduced as a result of segregation. As of yet we only have a partial understanding (of what causes this). It is a matter of movement behaviour being influenced by many of a particle's properties such as its size, shape, density, surface roughness, forces of attraction, friction etc. An additional factor is the number of industrial mixers with their own specific flow

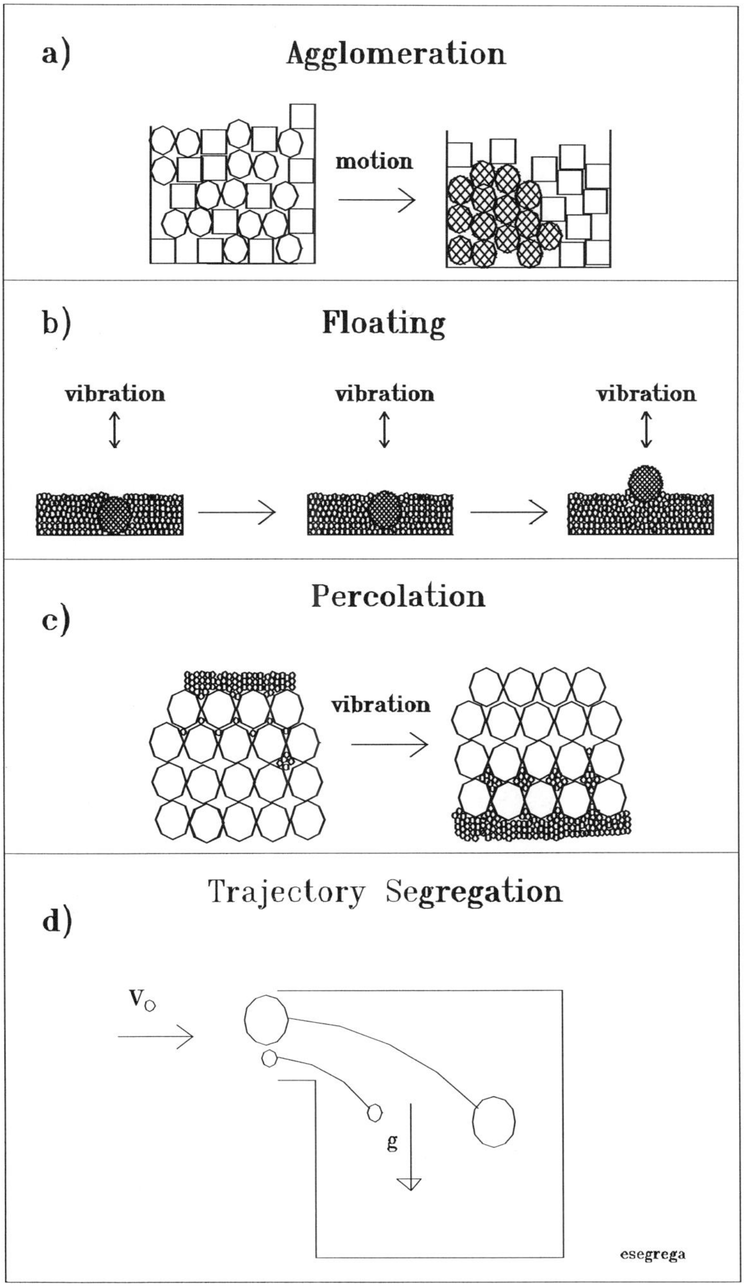

Fig. 2.2 Diagrammatic illustration of the mechanism of segregation

conditions. According to Williams [3] the particle's size is, however, the dominant influence in segregation. Since there is a divergence of particle sizes in nearly all industrial powders in such a way that they can be considered as solids mixtures of particles of different sizes, segregation is one of the characteristic problems of solids processing.
If mixtures are unsuitably transported, or if they are stored or moved in some way they will separate according to particle size and thus segregate. Figure 2.2 shows a schematic of the mechanism of segregation.
Figure 2.2a shows how segregation arises through the agglomeration of one component in 2-ingredient mixture. Agglomerates form when there are strong inter-particle forces and for these forces to have an effect the particles must be brought into close contact. In the case of agglomerates the particles stick to each other as a result, for example, of liquid bridges formed in solids, if a small quantity of fluid is present. Electrostatic forces likewise see to the cohesion of agglomerates. Van der Waals forces, reciprocal induced and dipolar, operate particularly upon finer grains smaller than 30 µm and bind them together. Shear forces must be employed during mixing to break up these agglomerates and high speed impellers or knives, for example, are used in the mixing chamber to do this.
Agglomeration can however have a positive effect on mixing. If a solids mix contains a very fine ingredient with particles in the submicron range (e.g. pigments), these fine particles coat the coarser ones. A mixture occurs which is stabilised by the van der Waals forces and is thereby protected from segregation.

If a solids mix is *vibrated*, the coarser particles *float up* towards the gravity force and collect near the open surface area. Figure 2.2b illustrates this effect in the case of a large particle which is lifted up slightly from the floor of the hopper by vibration. Smaller particles flow into the vacant space, preventing the larger from reclaiming its original position. If the large pellet has a high density it will compact the smaller ones when it falls down, further reducing their mobility. Solely because of the blocking effect of the larger particle's geometry there is little probability that this effect will run in reverse and that a bigger particle will take over the place left by a smaller one which has been lifted up. The large particle would also in this case have to displace several smaller ones. As a result the propability is higher that the coarse particles climbs further up.

Percolation (Fig. 2.2c) is by far the most important segregational effect. Finer particles trickle down through the gaps between the larger ones. These gaps act like a sieve. If a solids mixture is moved, gaps briefly open up between the grains and are picked out and passed through by the finer particles. Granted a single layer has a low degree of separation, but a bed of powder consists of many layers and interconnecting grades of particles which then together produce a significant divisions between fine and coarse grains (see fig. 2.3.) The result of percolation is a widespread segregation and percolation occurs even where there is but a small difference in the size of the particles. [3, 52].

The most significant economical example is the poured heap appearing, for example, when filling and discharging bunkers. A mobile layer with a high speed gradient forms on the surface of such a cone, which, like a sieve, bars larger particles from passing into the cone's core. Large grains on the cone's mantle obviously slide or roll downwards. But large, poorly mixed areas occur even inside the cone. Thus filling a silo or emptying it from a central discharge point is particularly critical. Segregation can be avoided if a mass flow can be achieved, i.e. the silo's content's move downwards in blocks.
Liquids can be efficiently blended by shaking them. With solids on the other hand, shaking promotes percolation and leads to segregation [3].

Segregation in the flow
This encompasses several effects which share the common factor of a gas contributing to the segregation processes. Here we will only go into segregation in solids. Naturally, segregation also occurs in the gas and solids phases as it is a forced requirement with cyclones and which has the effect of separating gases and solids in a gas/solids suspension so that the solids are left in the form of e.g. strands or aggregates [34].

Fig. 2.3 Pouring a heap - segregation caused by percolation
top left: pouring a heap by discharging from a belt top right: the coarse particles slide downward the cone bottom: accumulation of the coarse at the base of the heap Videoprints derived from: BASF- Product Properties and Process Engineering-Video Tutorial [49]

Figure 2.2d shows two particles of different sizes and density blown into a silo horizontally at a speed v_0. The forces of weight and resistance, which is a function of the relative speed squared and the drag coefficient c_w, affect the particles in the static gas atmosphere. For laminar flow around the particles the traversed path s is calculated using the equation shown in 2.1.

$$s = \frac{v_0 \rho_s d_p^2}{18\eta} \tag{2.1}$$

The particle size d_p enter the flight course squared, i.e. twice as large a particle covers four times the horizontal path. This likewise leads to a separation of the blown-in mixture.

The above equation applies to particles in an ideal flow pattern with very low concentration of the solids. The whole process becomes considerably more complex if one no longer takes the case of an individual pellet nicely dispersed in a gas but that of particles interacting within the stream due to high concentration of the flowing suspension [52]. If solids are poured out into a silo a circulating current of gas is created into which fine particles are sucked in.

Fig. 2.4 Banded segregation in a drum caused by differences in particle size
top: The components are quite well premixed before the mixer is started.
bottom left: The mixers rotates slowly, no cascading effect; after some time banded segregation occurs **bottom right:** segregation completed
Videoprints derived from: BASF- Product Properties and Process Engineering - Video Tutorial [49]

The videoprints [49] in figure 2.4 show an extreme example for segregation. A plexiglas drum is filled with two fractions of monosized Polyethylene bills, which differ in colour (red and blue) and silghty in size. Before the mixer is started the mixture is already quite well premixed (fig.2.4 top).
With increased mixing time, however areas of one or several ingredients form as stripes running perpendicular to the axis of mixing (fig 2.4 bottom) after finally the mixture is nearly completely segregated. The drum in fig 2.4 rotates slowly with very limited convective action. The particles roll on each other and „feel“ their differences in size. Industrial examples of mixers with similar mixing effect are tumble mixer and V-blenders.
This description of phenomena does not make any claims to being exhaustive and

segregation processes are not yet fully understood. Williams gives an overview of the literature on the subject [3] and suggests the following measures to counter segregation:

- The addition of a small quantity of water forms water bridges between the particles, reducing their mobility and thus stabilising the condition of the mixture. Because of the cohesive behaviour of particles smaller than 30 μm (ρ_s=2-3 kg/l) the tendency to segregate decreases below this grain size.
- Inclined planes down which the particles can roll should be avoided.
- In general, having ingredients of a uniform grain size is an advantage in blending.

3 The Statistical Definition of Homogeneity

For judging the efficiency of a solids blender or in general, of a mixing process, the status of mixing has to be quantified, thus a *degree of mixing* has to be defined. Here one has to specify what *property* characterises a mixture. This property is a product of a mixing process's end goal and which, for example, should be that of arriving at an even distribution of the ingredients or a uniform temperature throughout the volume of material. There are circumstances where a good mix can be called for in respect of several properties, e.g. particle size and composition.
As already discussed in the introduction, the volume or quantity wherein the mix should be at its most homogeneity is an essential element in evaluating a solids mixture. The mixture's condition is traditionally checked by taking a number of samples with this self same volume. These samples are then examined to see how widely that property is dispersed between them.

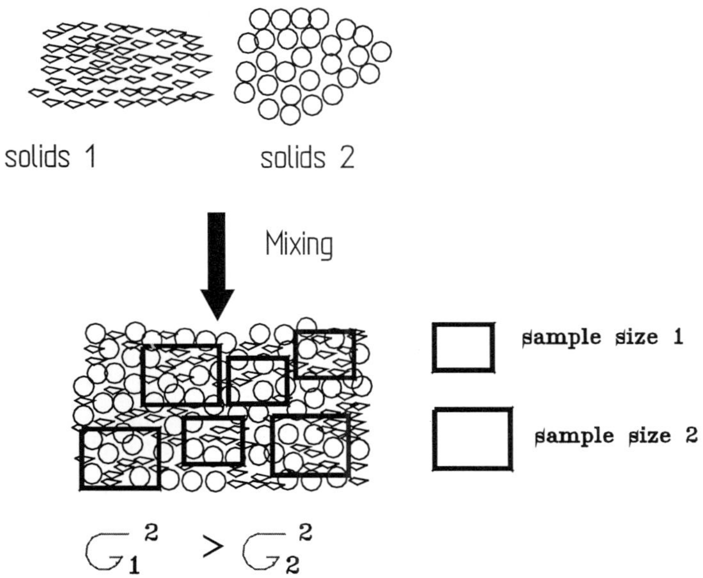

Fig. 3.1 The influence of the size of the sample on the numerical value of the degree of mixing

The *size of the sample* thus represents the resolution by which a mixture can be judged. The smaller the size of sample the more closely the condition of the mixture will be scrutinised (see Fig. 3.1). Dankwerts terms this the *"scale of scrutiny"* [4].

Specifying the size of the sample is therefore an essential step in analysing a mixture's quality, since it quantifies the mixing task from the outset. The size of the sample can only be meaningfully specified in connection with the mixture's further application. In a pharmaceutical production there has got to be a guarantee that active ingredients are equally distributed e.g. within the individual tablets in a production batch. So the size of the sample for testing the condition of a mixture is one tablet. This is less critical in the base materials industry where the sample size can even be in tons. However the correct sample size is less obvious in many processes using solids. Here is an example:

Silicon carbide is produced in an electric resistance furnace (an Acheson furnace) from silica sand and petroleum coke [5]. The reaction is highly endothermic.

$$SiO_2 + C \rightarrow SiO + CO$$

$$SiO + 2C \rightarrow SiC + CO$$

secondary reactions:

$$SiO + C \rightarrow Si_g + CO$$

$$SiC + 2SiO_2 \rightarrow 3SiO + CO$$

$$C + Si_g \rightarrow SiC$$

Besides silica sand with a typical particle size of 0.5 mm, the base material used in forming a furnace charge is petroleum coke with an average particle size of 3 mm. According to Babl and Steiner [5] the raw ingredients must be *"well mixed"*. Modern furnaces have a capacity of 400 tons and their warm up and reaction time is about 40 hours. Higher educt turnover can only be achieved when small quantities of both partners in the reaction and the surrounding gaseous ingredients are mixed together. Thus gas diffusion is able to equalise any local segregation. The clear difference in the main ingredients' particle size will show up in a difference between the particles' movement, which can lead to the components segregating when the furnace is being charged. On the other hand however, a low SiC yield, i.e. low reactor efficiency, cannot automatically be blamed on a problem with the mix since this has other possible causes.

Samples should be taken from various points in the furnace when checking the bulk's homogeneity within it. This in turn poses the question of the size of the sample. How large can the maximum area of segregation be without impeding the reaction process? The answer to this question calls for a detailed knowledge of chemical and physical processes at work during the reaction. The reaction contains several phases and complex heat conductance and material transition processes. A series of laboratory tests in which both components, the silica sand and the petroleum coke, are interlayered could be informative (Fig. 3.2).

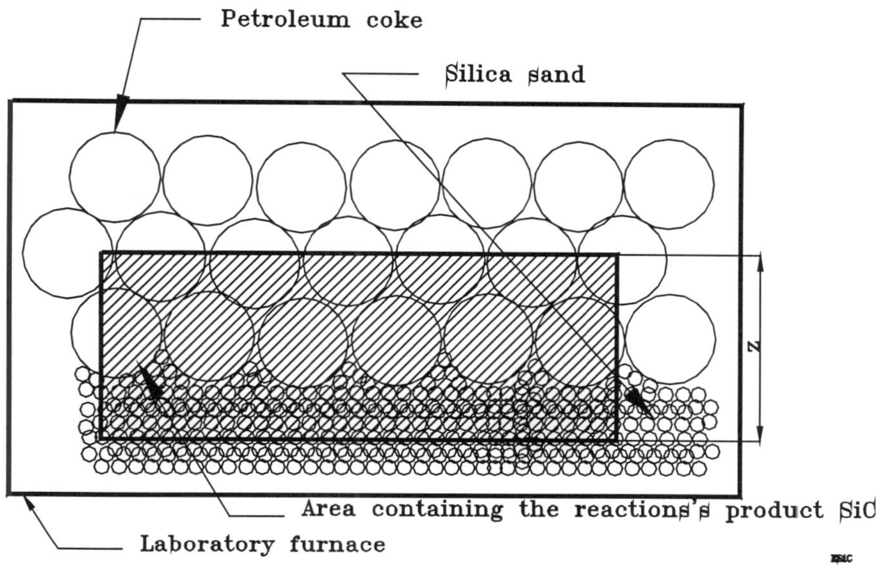

Fig. 3.2 Experimentally determining the volume of the sample $z \bullet z \bullet z$ in the reactant silica sand and petroleum coke mix

Since both ingredients are present in a segregated form there will only be a reaction and therefore conversion of the educts into SiC in a narrow band extending to a distance *z* vertical to the parting plane. Carried over into the three dimensional processes in the production furnace the volume of the sample should be z^3. This example is to show that in many cases the relevant sample sizes have to be established by experiments or by using calculation models. Once this is done you can begin a quality analysis of the mixture.

The traditional and general procedure is to take *identically sized* samples of the mixture from various points *at random* and to analyse them (in an off-line analysis). In the above example of producing silicon carbide one can use a 2-element mixture as a starting point for the sake of simplicity. But multi-element mixtures can also be described as twin ingredient mixes when a particularly important ingredient, e.g. the active agent in pharmaceutical products, is viewed as a tracer element and all the other constituents are combined into one common ingredient. This is a simplification of the statistical description of solids mixtures. When 2-element mixtures are being examined it is sufficient to just trace the concentration path of just one ingredient, the tracer. There will be a complementary concentration of the other ingredients. The description is completely analogous when the property or characteristic feature in which we are interested is not the concentration but e.g. moisture, temperature or the particle's shape.

If the tracer's concentration in the mixture is p and that of the other ingredients is q we get the following relationship:

$$p + q = 1 \qquad (3.1)$$

If you take samples of a specified size from the mixture and analyse them for their content of the tracer, the concentration x_i in the samples will fluctuate randomly around that the tracer's concentration p in the whole mixture (the "base whole"). *Therefore a mixture's quality can only be described using statistical means.* The smaller the fluctuations in the samples' concentration x_i around the mixture's concentration p the better its quality. *Variance* is a mathematical value frequently used in describing this relationship and which consequently is frequently defined as the *degree of mixing*.

3.1 Variance as a Measure for the Degree of Mixing

As with other quality controls a statistical interpretation of the results is required when analysing the quality of a mix. This sometimes presents difficulties for engineers in whose training both random processes and statistics were for the most part only used in a very scanty form. So it is not surprising that a heat exchanger is generally better evaluated than a mixer since the customer is able to arrive at an unambiguous specification. However, nonsensical requirements as regards mixture quality are often set in the case of mixers.

The less the local (in the case of batch mixing) or temporal deviation (with continuous mixing) of the whole mix's concentration p the higher the mixture's quality. For example, in a 2-element mix this is defined mathematically using the *concentration's variance* σ^2. For that reason the *variance* is frequently defined as degree of mixing. There are many more definitions of mix quality in literature on the subject but in most instances these relate to an initial or final variance and are frequently too complicated for industrial application [50]. The theoretical variance for finite sample numbers is calculated as follows:

$$\sigma^2 = \frac{1}{N_g} \sum_{i=1}^{N_g} (x_i - p)^2 \tag{3.2}$$

The variance is obtained by dividing up the whole mix, the base whole, into N_g samples of the same size and finding the concentration x_i in each sample.

If one analyses not the whole mix but a number *n* of *randomly* distributed samples across the base whole you will get the " *sample variance*" S^2 [1]. If this procedure is repeated several times a new value for the sample variance will be produced on each occasion. It is a statistical value. Thus each S^2 represents an estimated value for the unknown variance σ^2. In many cases the concentration p is likewise unknown and the random sample variance is then defined using the *arithmetical average* μ of the sample's concentration x_i.

---

1 In addition to mixture quality the random sample variance also enables a confidence interval to be calculated for the concentration p using Student's distribution.

$$S^2 = \frac{1}{n-1}\sum_{i=1}^{n}(x_i - \mu)^2 \; ; \; \mu = \frac{1}{n}\sum_{i=1}^{n} x_i \tag{3.3}$$

The random sample variance data makes little sense if you do not state how accurately it describes the unknown, true variance σ^2. The desired value is therefore the data in a confidence interval for σ^2. The confidence interval is the range of values within which there is a certain probability that varianceσ^2 is to be found. The distribution of S^2 is required in order to calculate the confidence interval. From the central set of limit values it can be deduced that the concentration in the samples will be normally distributed in an approximative fashion [6]. It is also known that the random sample variance will then possess a distribution χ^2 with a degree of freedom n-1, whereby the expected value of the random sample variance corresponds to the true variance:

$$E\{S^2\} = \sigma^2 \tag{3.4}$$

The probability W that the variance lies in a specific confidence interval can be calculated using the following equation [7]:

$$W\left((n-1)\frac{S^2}{\chi_u^2} < \sigma^2 < (n-1)\frac{S^2}{\chi_l^2}\right) = \Phi(\chi_u^2) - \Phi(\chi_l^2) \tag{3.5}$$

$\Phi(\chi^2)$ is the summation of the χ^2 distribution. χ^2_u and χ^2_l designate the upper and lower limits respectively. $\Phi(\chi^2)$ is tabled in statistical text books. The confidence interval used in mixing is mostly a unilateral one. Interest is focussed on the upper limit, which, with a given degree of probability, will not be exceeded by the variance. Raasch and Sommer [7] show that this requirement alters the equation as shown in 3.6 below:

$$W\left(\sigma^2 < (n-1)\frac{S^2}{\chi_l^2}\right) = 1 - \Phi(\chi_l^2) \tag{3.6}$$

The following example shows how this can be demonstrated:
n=101 samples of the same size are taken from a mixture. The random sample variance is calculated as per equation 3.3. The figure for freedom of movement is n-1=100. A figure of 95% is set as being the probability that the variance will be smaller than the upper confidence limit. Thus $\Phi(\chi^2_l) = 0.05$. The table for the summation function $\Phi[0.05: n-1]$ gives a value of 77.9 for χ^2_l.

Table 3.1 Values of the summation function χ_l^2 (0,05) for a probability of 95% [as per Kreyszig, 8]

(n-1)	5	10	20	30	40	50	70	100
χ_l^2	1,15	3,94	10,85	18,5	26,5	34,8	51,7	77,9

If you insert these values into equation 3.6 they produce the following. At a probability of 95% the mix's variance σ^2 is less than 1.28 times the measured sample variance S^2. This may astonish you. In spite of the large number of samples the upper confidence limit is still 28% above the experimentally determined value for S^2. For 11 samples (n = 11) the upper limit is even 2.5 times the sample variance.

Fig. 3.3 shows the size of the confidence interval normalised with the sample variance over the number of random samples n. The confidence interval depicts the accuracy of the analysis. The smaller the interval the more exactly the mix quality can be estimated from the measured sample variance. If there are few samples the mix quality's confidence interval is very large. An evaluation of the mix quality with a high degree of accuracy (a small confidence interval) calls for a lot of samples. This makes mix quality analysis expensive and requiring a great deal of effort.

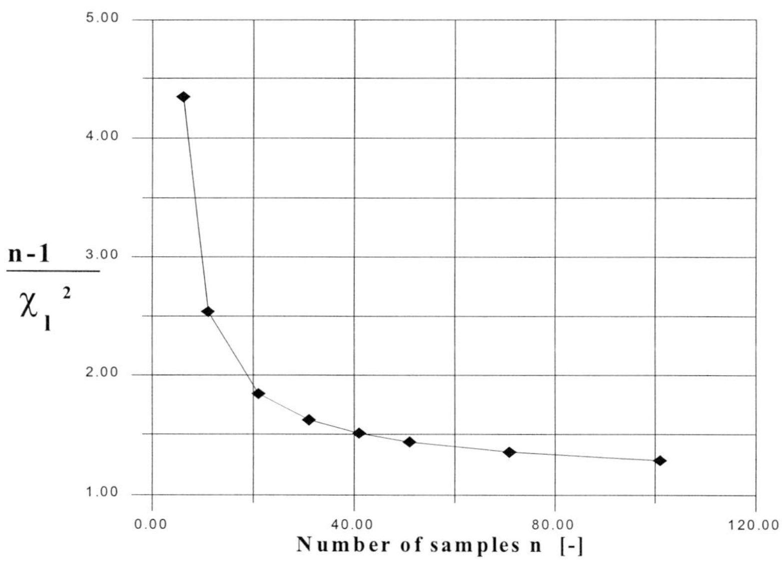

Fig. 3.3 The size of the unilateral confidence interval (95%) as a function of the number n of samples taken, measured in multiples of S^2 (cf. equation 3.6). Example: if 31 samples are taken the upper limit of the variance's confidence interval assumes a value 1.6 times that of the experimental sample variance S^2.

3.2 Ideal Mixtures

A *perfect mixture* [9] is when the concentration at any randomly selected point in the mix in a sample of whatever size is the same as that of the overall concentration. The variance of a perfect mixture has a value of zero. This is only possible with gases and liquids which can be mixed molecularly and where samples of the mixture are many times larger than its ingredients, the molecules. However, as soon as a detection system for gas mixtures exists with which mix conditions can be ascertained on a molecular level the concentration in samples will fluctuate randomly even in respect of such mixes.
In the case of solids mixes the intensity of the particles vis a vis the size of the sample or the sensor area must be taken into account. σ^2 thus depends on the size of the sample. There are two limit conditions of maximum homogeneity which are the equivalent of a minimum variance: an *ordered* and a *random* mixture. Fig 1.2 (in Ch. 1) shows a diagrammatic illustration of a 2-ingredient mix. The particles of one solid are represented by white rectangles and those of the other by black ones. As opposed to fig. 1.2, real solids mixes can consist of more than two ingredients and likewise the particles can differ distinctly both in shape and size.

Ordered mixes

The components align themselves according to a defined pattern. Whether this ever happens in practice is debatable. There exists the notion that because of interparticle processes of attraction this mix condition can be achieved. The interparticle forces find themselves in an interplay with those of gravity, which would prevent this type of ordered mix in the case of coarser particles. Interparticle forces predominate in the case of finer particles, i.e. cohesive powders. Ordered agglomerates or layered particles can arise. Sometimes not only the mix condition but also the mixing of powders in which these forces of attraction are significant is termed "ordered mixing" [11]. However Egermann [12] points to the fact that one should only use "ordered mixing" to describe the condition and not the mixing of fine particles using powerful interparticle forces. To achieve an ordered mix condition there must be a regular, ordered adhesion of the fine and the coarser ingredients. A mixing process is always one which creates disorder and only randomly structured adhesion can be achieved. Corresponding publications taking the opposite view have in the meantime turned out to be in error. Therefore according to Egermann's interpretation a clear distinction must be drawn between the ordered mixture condition and the mixing of powders using interparticle forces.

Random mixes

A random mix also represents an ideal condition. Raasch [6] defines random mixes as follows:

A uniform random mix occurs when the probability of coming across an ingredient of the mix in any subsection of the area being examined is equally high at any point in time for all subsections of the same size.

This is on condition that the particles can move freely. According to Lacy [13], *random mixing* is calculated as follows for a 2-ingredient blend in which the particles are of the same size:

$$\sigma^2 = \frac{p \cdot q}{n_p} \tag{3.7}$$

p stands for the concentration of one of the ingredients in the mix, q for that of the other (q=1-p) and n_p for the number of particles in the sample. It should be noted that the variance of the random mix grows if the number of samples decreases. The variance for *a completely segregated system* is calculated as shown in equation 3.8:

$$\sigma^2_{segregated} = p \cdot q \tag{3.8}$$

Equation 3.7 is a highly simplified model, for no actual mixture consists of particles of the same size. It is likewise a practical disadvantage that the number of particles in the sample has to be known in order to calculate the ideal mix as per equation 3.7, however the sample volume is usually specified. In 1954 Stange [14] calculated the variance of a random mix in which the ingredients possess a distribution of particle sizes. He made an allowance for the average mass m_p and m_q of the particles in each component and the particle mass's standard deviationσ_p and σ_q. These values can be drawn from the numerical distribution[2]. He designates the "variability" c to the standard deviation and average particle weight quotients:

$$c_p = \frac{\sigma_p}{m_p} \quad ; \quad c_q = \frac{\sigma_q}{m_q} \tag{3.9}$$

Variability is a measure for the width of the particle size distribution. The higher the value of c the broader the particle size distribution. The size of the sample is now specified in practice by the mass M and no longer by the number of particles n_p as shown in equation 3.7. Stange gives the following relationship for the variance in random mixes:

$$\sigma^2 = \frac{pq}{M}\left[pm_q\left(1+c_q^{\ 2}\right)+qm_p\left(1+c_p^{\ 2}\right)\right] \tag{3.10}$$

---

[2] see also the example in Chap. 4

If the components have a narrow grain size spectrum, i.e. c_q and c_p are low, the random mix's variance falls. The same is obtained by pulverising the particles since the mean grain masses decrease. As in equation 3.7 the variance in a random mix is also inversely proportional to the size of the sample. Equation 3.8 for calculating the variance in a fully segregated system also applies when p and q are defined in terms of mass concentration.

When calculating the *variance of an ideal random mix* Sommer [15] takes account of the ingredients' actual particle size distribution. He distinguishes between two cases:

1) Both ingredients of the mix possess the same particle size distribution:

$$\sigma^2(x) = \frac{p}{M}\int_0^1 g(\hat{R})d\hat{R} - \frac{2p^2}{M}\int_0^1 (1-\hat{R})\cdot\int_0^{\hat{R}} \frac{g(\varsigma)}{(1-\varsigma)^2}d\varsigma d\hat{R} \tag{3.11}$$

"p" stands for the average tracer element content, "M" for the sample's volume, "$\hat{R}$" the residue and "$g(\hat{R})$" the relationship of a particle's volume to the residue. This is calculated from the particle size distribution by converting the particle diameter into a particle volume.

2) One component is clearly larger than the other and their distribution does not overlap:

According to Sommer the random mix variance is calculated as per equation 3.12:

$$\sigma^2(x) = \frac{q^2 p}{M}\cdot\int_0^1 \frac{g(Q_3^c)}{(q - p\cdot Q_3^c)^2}\, dQ_3^c \tag{3.12}$$

"c" is the index for the larger particles and "Q_3^c" the grain size transient distribution.

3.3 Scale and Intensity of Segregation

3.3.1 Time domain analysis using the autocovariance function

The account in the previous chapter of how to ascertain the quality of a mix using its variance represents the traditional method of analysing the mixing process. The size of samples and what is to be taken as the reliable variance are specified according to the mix's application. Samples are taken at random from the flow of material or from the mixer. This minimises the still considerable number of samples required. **Random** sampling in particular is very expensive. For example, no one spot should be preferred from which to take samples since that could lead to a drastic distortion of the mix's analysis. This is avoided, for example, by dividing up the mix area into numerous consecutively numbered cells. A small section of these cells is picked out using random numbers. Samples are then taken from those randomly selected cells and the variance established.

Frequently however, **systematic** instead of random sampling is chosen, as illustrated in Fig. 3.4. What is meant by systematic sampling is that the samples are taken at regular points in time or local spatial intervals:

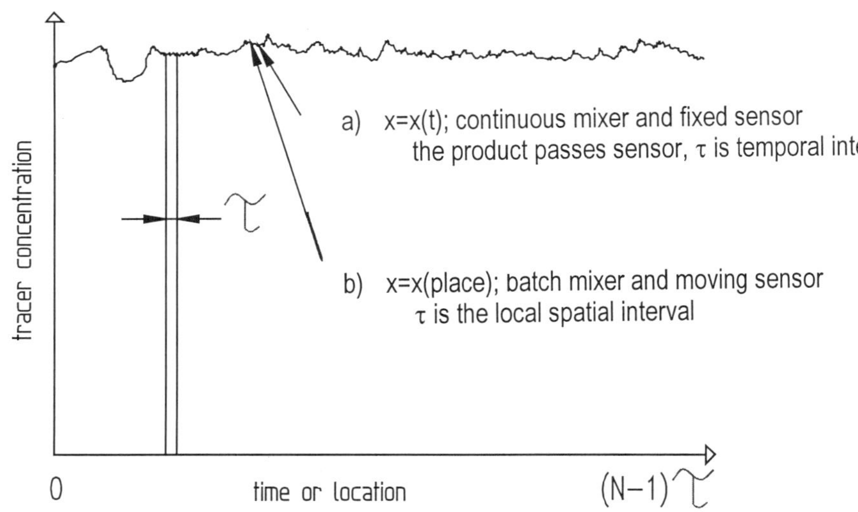

Fig. 3.4 Systematic sampling: Concentration as a function of time or location

Particular care needs to be taken when analysing these systematic measurements. If solely the variance is to be considered in assessing a mix's condition this can lead to false readings if the interval τ between the samples or measurements coincides with the cycle of an existing fluctuation in the process. Therefore in the case of systematic measurements one must try to increase the number of samples so that the mixing process is described by a dense matrix of measurement points. This can only be achieved using on-line measurement whereby sampling and analysis is automated and the mixing process can be followed with a high degree of temporal and spatial resolution. We know from measurement and control technology that a concentration's course can only be fully recorded if the on-line measurement sampling frequency is at least twice as high as the highest frequency occurring in the process (Shannon's sampling theorem).

At present there unfortunately is no on-line measurement system which can provide this type of checking for the multitude of industrial mixing processes. On-line monitoring is however possible for certain mixture models and an example is presented in Chapter 5.

Stange [16 & 15] is the first to analyse systematic measurements and compare errors arising from random and systematic sampling. This is shown in an example whereby a coal's ash content at the end of a processing plant is established.

Although the samples are taken manually at long intervals in time (samples are taken from the conveyor belt flow) he shows that you have to know the correlation or relationship between them, the autocorrelogram. This particularly affects measurėments taken using an on-line system.

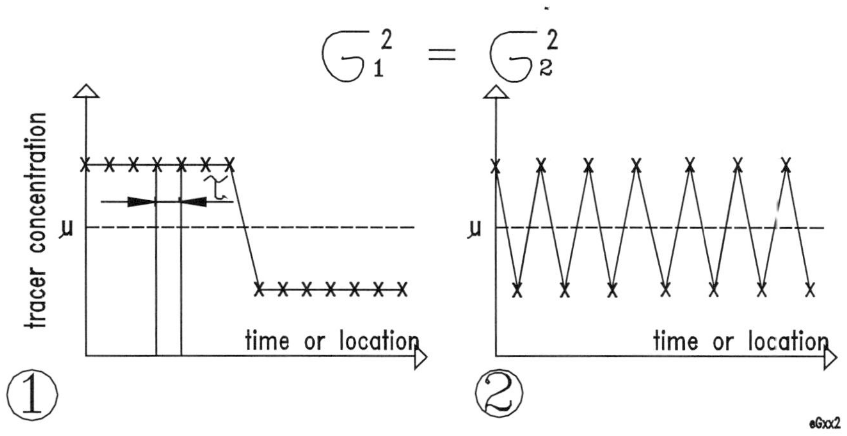

Fig. 3.5 The variance's dual significance when specifying the quality of a powder mix

The variance reading is not enough for analysing systematic measurements. In general it has the following disadvantages when analysing mixing processes.

- It does not contain any data on the mix's structure. Fig. 3.5 shows two mixes which in fact present the same variance but have a differing mixture structure. It does however correctly report the concentration's fluctuation scatter around the average value and this is sufficient for many applications.
- The variance changes with the size of sample. Only with random or completely segregated mixes is the functional relationship known between the variance and the sample size. In the first case the variance is inversely proportionally to the sample size and in the second it remains unaffected by it (cf. equations 3.7 and 3.8).

Dankwerts [4] puts forward the proposal that a mixture's homogeneity is described using two values: the "**scale of segregation**" and the "**intensity of segregation**". The latter designates the variance. The concept behind the scale of segregation is a simple one. If you take samples from a mixture at a set temporal or spatial interval τ a correlation will exist between neighbouring samples and they are "related". This "relationship" decreases as the interval increases provided there has been no fluctuation or large scale separation of the mix's components. The relationship is expressed mathematically by what is termed an *autocorrelogram.* The characterisation of solids mixes using autocorrelograms and associated applied statistical functions is discussed in the following.

It is assumed that there is a concentration cycle over the time (time series). This type of concentration cycle as shown in Fig. 3.4 can for example be the result produced by an on-line measurement at a continuous mixer's output or systematic sampling from the conveyor belt. The following analysis presented also applies if the concentration course is not given as a function of time (time series) but of location.

We are assuming that the mixing process is approximately stationary[3]. Stationary means that in the case of continuous mixing all relevant values such as the filling ratio or the feeder flow rate are also stationary. In batch mixing all start-up processes must already have been concluded. The concentration course can then be described statistically using the **autocovariance function $C_{xx}(\tau)$** [17]:

$$C_{xx}(\tau) = \lim_{T \to \infty} \frac{1}{T} \int_{t=0}^{T} \{x(t) - \mu\}\{x(t+\tau) - \mu\} dt \tag{3.13}$$

It contains the temporal interval τ as a parameter. The autocovariance function has meanwhile been included into many scientific computer programs. Today the difficulty only lies in getting this quantity of systematic measurement data to give a meaningful calculation of the autocovariance function. In Stange and Dankwerts' time, calculation of the autocovariance or the autocorrelogram derived from it was very labourious and time consuming. This is certainly a reason for the scale of segregation's limited adoption. The autocovariance function is closely related to the **autocorrelation function $R_{xx}(\tau)$**. The difference between them is the additive term μ^2, the squared average value of the tracer concentration in the samples:

$$C_{XX}(\tau) = R_{XX}(\tau) - \mu^2 \tag{3.14}$$

The **autocorrelogram** $\rho_{xx}(\tau)$ is the autocovariance function normalised by the variance:

$$\boxed{\rho_{XX}(\tau) = \frac{C_{XX}(\tau)}{\sigma^2}} \tag{3.15}$$

The sum of the autocorrelogram is always less than or equal to 1:

$$|\rho_{xx}(\tau)| \leq 1 \tag{3.16}$$

If there is a maximum "relationship" between samples at an interval τ, $\rho_{xx}(\tau)$ equals 1. In a random process such as a random mix there is no relationship between the measurements or samples and, $\rho_{xx}(t)$ becomes zero for all values of interval τ. A

---

[3] The analysis applies to stationary, ergodic random processes

random mix's autocorrelogram only reaches a value of 1 at a zero interval, i.e. simultaneousness or identicalness. Such a course is also termed the delta function (see Fig. 3.6 for C_{xx}). In reality, mixes lie between these two extremes. If the interval τ increases, the relationship and thus the correlation reduces and the mix becomes random. Bourne [18] and Stange [19] discuss models for an autocorrelogram's curve in solids mixes. Lacey and Mirza [20] have defined autocorrelograms for Y-mixers. Dankwerts [4] defines the **linear scale of segregation l_0** as areas subject to the autocorrelogram:

$$l_0 = \int_{\tau=0}^{\infty} \rho_{xx}(\tau)d\tau \tag{3.17}$$

l_0 contains the information on the mix's structure and is the measurement of the average size of the aggregates in the mix which contain too high or too low a concentration, thereby producing an elevated variance. l_0 also includes the interval unit of measurement, i.e. metres if measurements are taken spatially or seconds if temporally. In the latter case the flow passing the detector can be normalised.

$$l_0^* = \dot{m} \cdot l_0 \tag{3.18}$$

The characteristic conglomerate size is then given in kilograms.

The following also applies: *The variance σ^2 describes the segregations's intensity and the linear scale of segregation l_0 enables one to estimate the size of the segregation zone producing this increased variance.*

However a price has to be paid for such detailed definition. It takes even more random samples in order to determine the correlogram than are already needed to establish the variance. At the same time as Dankwerts, Stange [19] was putting forward similar reflections. He inferred the following relationship which lets one calculate the tracer component's concentration variance if the sample size (index N) exceeds a primary quantity (index 1) by N times. For this you need the tracer concentration's variance and autocorrelogram established from samples taken with a primary quantity (index 1) from the mix.

$$\sigma_N^2 = \frac{\sigma_1^2}{N}\left[1 + 2\sum_{i=1}^{N}\left(1 - \frac{i}{N}\right)\rho_{xx}(i)\right] \tag{3.19}$$

There are few measurement values for calculating mixture's correlogram of a real life powder although a mix is completely defined in detail using both its variance and correlogram or autocovariance function only. The aim of any test should be to collect numerous measurements, thereby making it possible to create an autocorrelogram. Fig. 3.6 is a collation of the statistical analysis of time series for analysing a mix. The closely related analysis of frequency range time series which will be the subject of the next section is also illustrated.

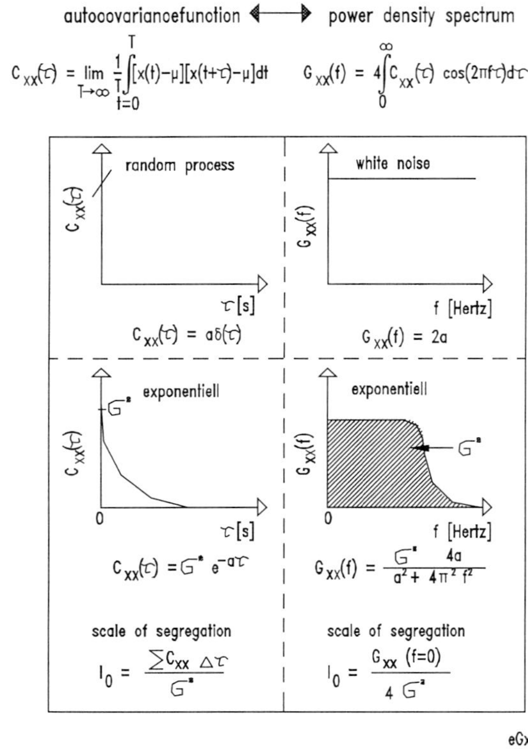

Fig. 3.6 Characterising a mix in the time domain and frequency range using the autocovariance function and power density spectrum

3.3.2 Frequency analysis using the power density spectrum

The time domain statistics just illustrated have a younger rival, that of statistical frequency range analysis. Frequency analysis is mainly used in measurement and control technology where the task is to statistically evaluate time series readings. In this field however the statistical analysis of assembled frequency range data (frequency analysis) is more often used than the autocovariance function when calculating a **power density spectrum**. This is particularly an option when the data has been recorded by a computer. The triumphal march of the power density spectrum formula has not halted since the Seventies when the fast Fourier transformation, a fast algorithm for calculating Fourier transformation, was developed. Like the autocovariance function it is also suited to analysing cyclic processes (see Ch. 8).

Frequency range analysis calls a mathematical operation, that of Fourier transformation of the data. Once the data has been Fourier transformed the tracer concentration is calculated in Bendat and Piersol's [17] discrete form as follows:

$$X_k = X(k\Delta f) = \Delta t \sum_{n=1}^{N} x_n \exp\left(-j2\pi \frac{kn}{N}\right) \qquad k = 1,2,3...N \tag{3.20}$$

$\Delta t = T/N$, T = measurement duration, N = number of measurements

According to Geering [21], there is a close relationship between the power density spectrum and autocovariance function. The unilateral power density spectrum G_{xx} is linked to the autocovariance function via Fourier transformation (see Fig. 3.6).

$$G_{xx}(f) = 4\int_0^\infty C_{xx}(\tau)\cos(2\pi f\tau)d\tau; \quad 0 \le f < \infty \tag{3.21}$$

G_{xx} is plotted across the frequency with the discrete frequency intervalΔf replacing gap interval τ:

$$\Delta f = 1/T \tag{3.22}$$

Turning this around, the autocovariance function can be obtained again from the power density spectrum's inverse Fourier transformation:

$$C_{xx}(\tau) = \int_0^\infty G_{xx}(f)\cos(2\pi f\tau)df \tag{3.23}$$

Incidentally, most computer programs nowadays calculate the autocovariance function via the power density spectrum followed by an inverse Fourier transformation.

It was shown in the previous section that the autocorrelogram for random mixes has a value of 1 where $\tau = 0$ and zero (delta function) for all other cases of t. It therefore follows from equations 3.15 and 3.13 that the autocovariance function C_{xx} of a random mix is also a delta function (see Fig. 3.6), where:

$$C_{xx}(\tau = 0) = \sigma^2 \quad C_{xx}(\tau \ne 0) = 0 \tag{3.24}$$

From equations 3.23 and 3.24 we get:

$$C_{xx}(\tau = 0) = \sigma^2 = \int_0^\infty G_{xx}(f)df \tag{3.25}$$

The area under the power density spectrum equals the variance. The power density spectrum can therefore also be interpreted as variance density distribution. Equation 3.21 applies in completely the same way:

$$G_{xx}(f = 0) = 4\int_0^{\infty} C_{xx}(\tau)d\tau \tag{3.26}$$

The last equation serves to produce the relationship between the **power density spectrum** and the **linear scale of segregation** l_o:

$$l_0 = \int_{\tau=0}^{\infty} \rho_{xx}(\tau)d\tau \tag{3.27}$$

$$\rho_{xx}(\tau) = \frac{C_{xx}(\tau)}{\sigma^2} \tag{3.28}$$

$$\Rightarrow l_0 = \frac{G_{xx}(f = 0)}{4 \cdot \sigma^2} \tag{3.29}$$

The linear scale of segregation l_o, which gives a feel for the size of the unmixed agglomerates in the mix, is determined from the power density spectrum with f=0 and the value $G_{xx}(0)$ being divided by four times the variance. The variance is thus the area under the power density spectrum (cf Fig. 3.6). As mentioned further back, a random mix has a delta function autocovariance where τ=0 and the area is σ^2. In the power density spectrum, random mixes are described by a line parallel to the abscissa (Fig. 3.6). This is termed *white noise* in control technology. Like white noise a random mix is a theoretical boundary case and does not exist in reality.

Example of a time series analysis using the power density spectrum

Silicon carbide is being mixed with aluminium hydroxide in a continuous mixer (a Gericke AG Multiflux). Both components are fed into the mixer at a constant flow rate. Two tests were conducted to check the effect of the positioning of tools in the Multiflux on the results from the mixer (see fig.3.7). The two configurations are designated Multiflux I and Multiflux II. Details of the tests are to be found in [30]. The silicon carbide's concentration in the output from the mixer is obtained every 0.2 seconds using a fibre optic probe (see Ch. 5). After 4096 measurements you get the time series illustrated in Fig. 3.8, i.e. the concentration of the tracer component SiC over the period.

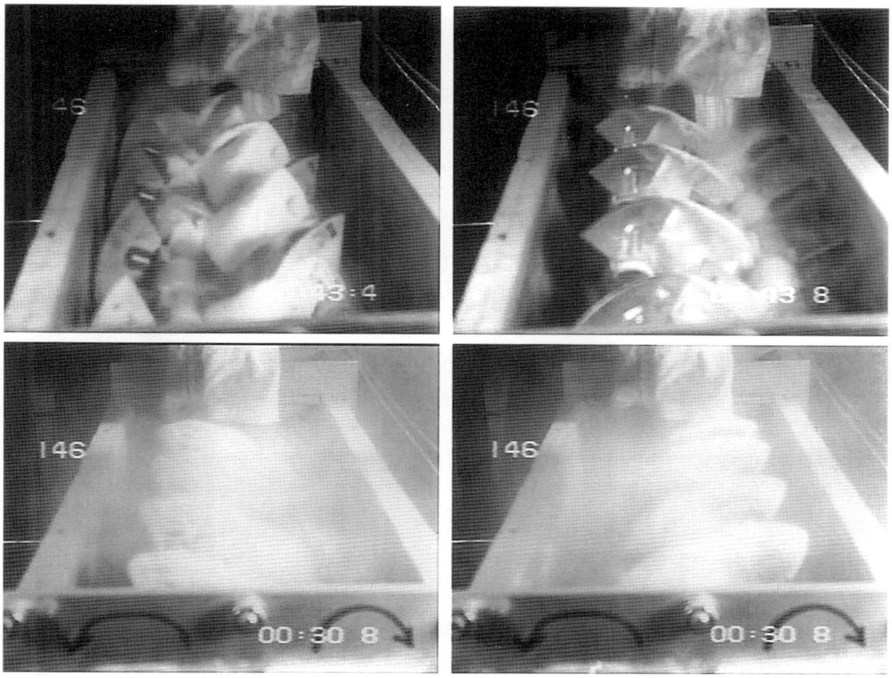

Fig. 3.7 Videoprints of the Multiflux-mixer showing the effect on positioning the tools [30]
top: Multiflux I bottom: Multiflux II

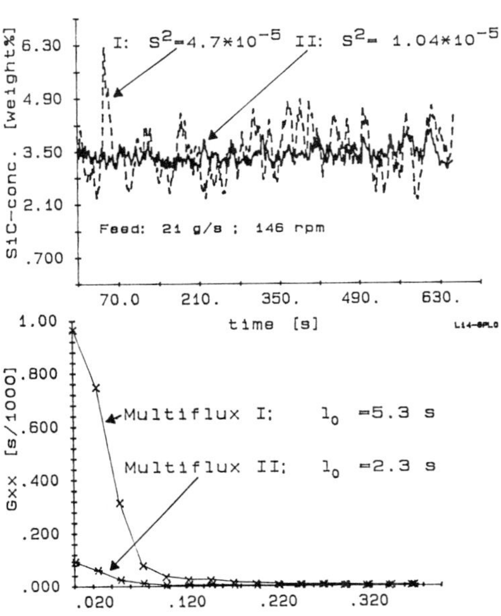

Fig. 3.8 Concentration fluctuations **a)** and power density spectra **b)** of the silicon carbide in the continuous Multiflux I and II mixers' output as obtained by a fibre optic probe.
Mix: SiC tracer ($d_{p50.3}$ = 26 μm) and $Al(OH)_3$ ($d_{p50.3}$ = 70 μm)

It is striking that the fluctuations in the tracer component's concentration with Multiflux II are distinctly lower. Weaker fluctuations and a smaller variance respectively demonstrate the better mixing effect using Multiflux II. The intensity of segregation is also low in comparison to that with Multiflux I.

Close scrutiny of both time series shows that the correlation of the concentration signals for Multiflux I continues for a longer measurement period. This is shown more clearly by the frequency analysis of both time series (Fig. 3.8b). The power density spectrum over the area corresponds to the variance already calculated. Here Multiflux I also produced a notably poorer mix with a higher variance. Furthermore, that spectrum is dominated by low frequencies whereas with Multiflux II the variance is evenly distributed across all the frequencies (and at a lower level). This is approaching the ideal random mix. If you calculate the*scale of segregation* l_o and the relevant *intensity of segregation* from the spectra (see equation 3.29) this gives 5.3 secs for Multiflux I against 2.3 secs for Multiflux II. Therefore Multiflux II clearly mixes more finely.

Determining the power density spectrum in practice

Computing power density spectra 25 years ago was still extremely calculation intensive. Since then the analysis of a time series using the power density spectrum or autocovariance functions has in the meantime been included in many scientific computer programs. Since the development of fast Fourier transformation a fast algorithm has been available and time series functions are mostly calculated from the power density spectrum. In spite of these good prerequisites the time series analysis applied in scientific publications frequently does not meet the minimum requirements. For the most part it presents a "raw"power density spectrum which has not undergone any further analysis. Which is why there follows a short guide to the practical determination of a power density spectrum. The starting point is the concentration's deviation from the average value for the sample:

$$\hat{x} = x - \mu \tag{3.30}$$

The following definition is used to determine the power density spectrum:

$$G_{xx}(f_k) = \lim_{T\to\infty} \frac{2\left|\hat{X}(k\Delta f)\right|^2}{T} \tag{3.31}$$

As with the variance, here too only an estimated value $\hat{G}_{xx}$ can be set for G_{xx} since an infinitely ongoing measurement is impossible. Like the variance, G_{xx} is alsoχ^2 distributed.

$$\hat{G}_{xx}(f_k) = \frac{2\left|\hat{X}(k\Delta f)\right|^2}{T} \quad ; \; k = 0,1,\ldots\ldots,\frac{N}{2}; \; f_k = k\Delta f \tag{3.32}$$

The estimate's accuracy can be improved upon by dividing up the measuring activity $\{x(t)\}$ over the period T_d into n_d sections of duration $T=T_d/n_d$. Each section contains N measurement points. The power density spectrum is calculated for each of these sections as shown in equation 3.32. These are then averaged out [17]:

$$\hat{G}_{xx}(f_k) = \frac{2}{n_d N \Delta t} \sum_{i=1}^{n_d} \left| \hat{X}_i(f_k) \right|^2 \tag{3.33}$$

As Bendat and Piersol [17] demonstrate, the following applies with a probability of 95%:

$$\left(1 - \frac{2}{\sqrt{n_d}}\right) \cdot \hat{G}_{xx}(f) \le G_{xx}(f) \le \left(1 + \frac{2}{\sqrt{n_d}}\right) \cdot \hat{G}_{xx}(f) \tag{3.34}$$

Equation 3.33 details a practical means of calculating the power density spectrum:

1) Measure the concentration over a period T_d
2) Divide this series of measurements into n_d identical portions T
3) Calculate the deviations from the average value (equation 3.30)
4) Fourier transform these portions, e.g. using a fast Fourier algorithm in a scientific software program (correct this if necessary with a Hanning Foster multiplication [17])
5) Calculate the unilateral power density spectrum (equation 3.33)

3.4 Mix Quality - a Summary

The mathematics presented in this chapter appear complicated at first glance. However, as with many things the concepts put forward will very quickly become clear through practice and usage. To end with we shall take another brief overview of this chapter (cf. Fig. 3.9):

Intensity of segregation

The quality of a mix is mostly judged by how much the tracer concentration in a number of samples deviates from its average concentration in the mix. This is expressed mathematically by the tracer concentration's *variance*. Variance is normally used to define a powder mix. It is the measure which enables the mix's quality to be determined using the smallest number of (random) samples. However, for the sake of randomness the minimum number of samples should be more than 25 with the result that the costs of conducting a mix quality analysis are still considerable. The random selection of samples likewise places high requirements in terms of the sampling. Random sampling and analysis using variance represents the standard procedure and can be widely applied.

Scale of segregation

If on the other hand samples are taken systematically at set intervals, and this is done in the case of all automated measuring methods (on-line, in-line or in situ), statements on the structure of a mix, the "linear scale of segregation", can be made. Calculation of the *autocovariance function* is required for this. Another description of the mix's condition is supplied by the power density spectrum, which, calculated using modern computer programs, very quickly provides that type of analysis. The relationship to methods hitherto employed in mixing has been shown and a specific definition of power density spectrum used to do so.
To define a mix's condition in the time domain you need the average value and autocovariance function (variance) and in the frequency range the average value and power density spectrum.

The power density spectrum like the autocovariance function are also suitable for cyclic processes. This will be described in Chapter 8. Both call for a comprehensive amount of data as can only be obtained using on-line testing. However in the case of solids systems in particular some development work has still to be done in that field.

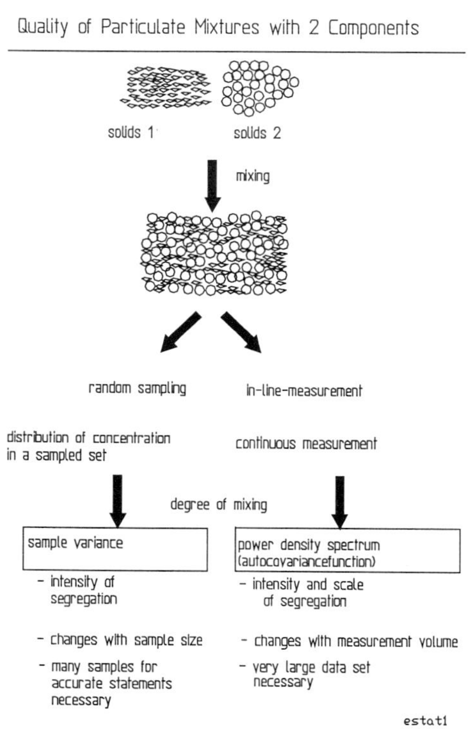

Fig. 3.9 Mix quality - a summary

4 Example: Mix Quality of a Sand and Cement Mixture

This chapter presents a mix quality analysis in the form of an exercise. This give the reader a chance to apply parts of the theoretically exacting statistical principles in the last chapter.

Three tons of a **sand** (80% by weight) and **cement** (20% by weight) mix have been produced. The quality of this mix has to be checked. 30 samples at 2 kg of the material mixture have been taken at random and the sand content in these samples established. The grain size of the sand and cement has been ascertained by sieve analyses.

Table 4.1 Grain size distribution of the sand and cement as found by sieve analyses

Cumulative frequency distribution of the sand (Transient): ρ_{sand}=2650 kg/m³		Cumulative frequency distribution of the cement (Transient): ρ_{cement}=3150 kg/m³	
Grain size class d_p, sand, i [mm]	Mass fraction dQ_{sand} [-]	Cement size class d_p, cement, i [mm]	Mass fraction dQ_{cement} [-]
0.1-0.5	0.01	< 0.01	0.01
0.5-1	0.15	0.01-0.025	0.1
1.0-3.0	0.4	0.025-0.04	0.29
3.0-7.0	0.44	0.04-0.09	0.55
		0.09-0.1	0.05
	Σ=1		Σ=1

The mass fraction of the sand x_i [kg_{sand}/kg_{mix}] in the samples comes to:
3 samples @ 0.75; 7 @ 0.77; 5 @ 0.79; 6 @ 0.81; 7 @ 0.83; 2 @ 0.85

Task: The *degree of mixing* defined as the variance of the mass fraction of sand in the mix needs to be determined. It has to be compared with the variance for a fully segregated system and the ideal variance of a random mix. What is the change in the ideal variance when the size of the samples is 1 or 10 kg respectively?

Solution:

4.1 Mix Quality

First of all the random sample variance S^2 is calculated and with it an upper limit for the true variance σ^2 can then be laid down (cf. equations 3.1 and 3.2). The sand's average concentration p in the whole 3 ton mix is estimated1 using the random sample average μ:

$$\mu = \frac{1}{n}\sum_{i=1}^{n} x_i = \frac{1}{30}\sum_{i=1}^{30} x_i = 0{,}797 \tag{4.1}$$

$$\begin{aligned} S^2 &= \frac{1}{n-1}\sum_{i=1}^{n}(x_i-\mu)^2 = \frac{1}{29}\sum_{1}^{30}(x_i-0{,}797)^2 \\ &= \frac{1}{29}\left(3\cdot 0{,}047^2 + 7\cdot 0{,}027^2 + 5\cdot 0{,}007^2 + 6\cdot 0{,}013^2 + 7\cdot 0{,}033^2 + 2\cdot 0{,}053^2\right) \\ &= 9{,}04\cdot 10^{-4} \end{aligned} \tag{4.2}$$

95% is set as the probability W determining the size of the confidence interval for the variance σ^2. An upper limit (unilateral confidence interval) is then calculated for variance σ^2 using equation 3.6:

$$W\left(\sigma^2 < (n-1)\frac{S^2}{\chi_l^2}\right) = 0{,}95 = 1-\Phi\left(\chi_l^2\right) \Rightarrow \Phi\left(\chi_l^2\right) = 0{,}05 \tag{4.3}$$

From the table of the χ^2 distribution summation function $\Phi(\chi_l^2; n-1)$ (Kreyszig's example table [8]) you get the following for 29 degrees of freedom at 17.7:

$$\sigma^2 < (n-1)\frac{S^2}{\chi_l^2} = 29\cdot\frac{9{,}04\cdot 10^{-4}}{17{,}7} = 14{,}8\cdot 10^{-4} \tag{4.4}$$

It can therefore be conclusively stated with a probability of 95% that the mix quality σ^2 is better (equals less) than $14.8\cdot 10^{-4}$. Note: the smaller the variance the lower the fluctuations and the better the mix quality.

Using this value for the variance and assuming a normal concentration distribution, a confidence interval for the concentration can now also be specified:

---

1 If the random sample variance is calculated using the true concentration p=0.8, n instead of n-1 will appear in the formula's denominator

$$\mu - \frac{t_{W,n-1} \cdot S}{n} \leq p \leq \mu + \frac{t_{w,n-1} \cdot S}{n} \tag{4.5}$$

$t_{w,n-1}$ is the "Student factor" and is presented in tabular form for various probabilities and degrees of freedom. 29 degrees of freedom and a probability of 95% gives $t_{95\%,29}$ at 2.04. The size of the confidence interval for the concentration p in the number of random samples n is determined in exactly the same way.

- *Variance of a completely segregated system:*

The variance of a completely segregated system can be calculated as a value for comparative purposes (cf. equation 3.8):

$$\sigma_{segregated} = p \cdot q = 0{,}8 \cdot 0{,}2 = 0.16 \tag{4.6}$$

This is the mix quality's worst case value (because it is the highest). (As a comparison: the mix has a quality of $14.8 \cdot 10^{-4}$.) Since, by way of an example, it describes the mix when the components are in a separated state before the actual mixing process the variance obtained after the mixing process is sometimes related to this initial value.

4.2 Stange's Variance for an Ideal Random Mix

According to Stange the mix quality for an ideal random mix can be calculated using the following formula (cf. equation 3.10):

$$\sigma^2 = \frac{pq}{M}\left[pm_q\left(1 + c_q^{\,2}\right) + qm_p\left(1 + c_p^{\,2}\right)\right] \tag{4.7}$$

with

$$c_p = \frac{\sigma_p}{m_p} \quad ; \quad c_q = \frac{\sigma_q}{m_q} \tag{4.8}$$

The variance coefficients c_p and c_q as too the average particle masses m_p and m_q represent values which have to be obtained from the numerical distribution Q_0. Given the fact that we have (through pass) sieve analyses of the sand and cement we therefore know the corresponding volume distribution Q_3. Using the moments method these distributions are intertransferable [65]. Stange himself used a closely related graphing method in his textbook "applied Statistics - Part 1" to calculate the products $m_q(1+c_q^{\,2})$ and $m_p(1+c_p^{\,2})$ from the sieve analyses' results. To this end the grain mass is plotted over the flow. Assuming the particles are spherical, grain masses are shown by the diameter of the particles using the following formula:

$$\hat{m}_{p,i} = \frac{\pi}{6}\rho_{Sand} \cdot (d_{p,Sand,i})^3 \quad \text{and} \quad \hat{m}_{q,i} = \frac{\pi}{6}\rho_{Zement} \cdot (d_{p,Zement,i})^3 \tag{4.9}$$

Figures 4.1-4.4 show the Q_3 distribution for sand and cement and the graphs obtained therefrom for the grain mass in question over the through pass.

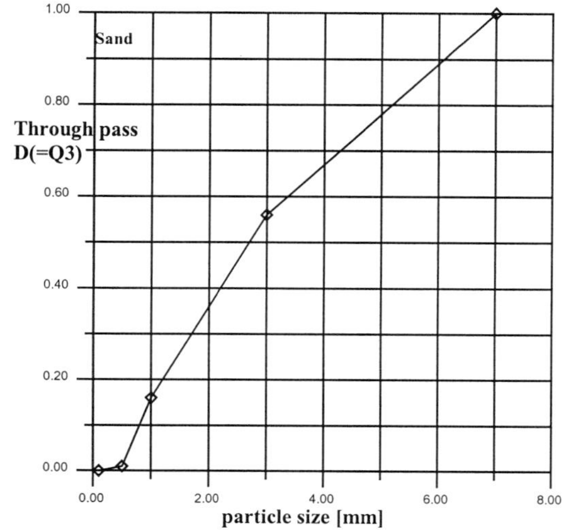

Fig. 4.1 Grain size of the sand - cumulative distribution (through pass) Q3

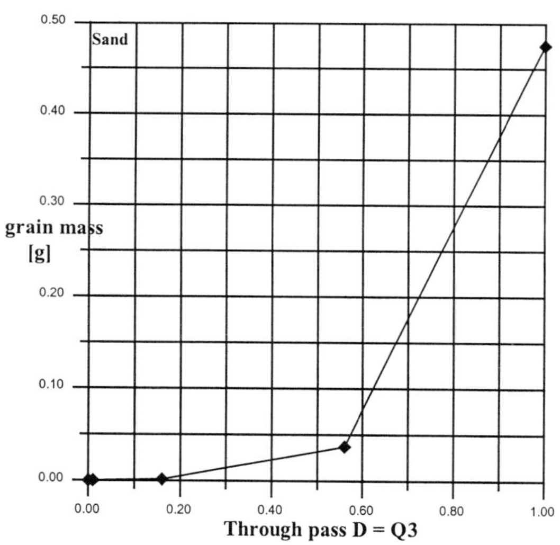

Fig. 4. 2 The sand's grain mass in relation to the through pass Q3

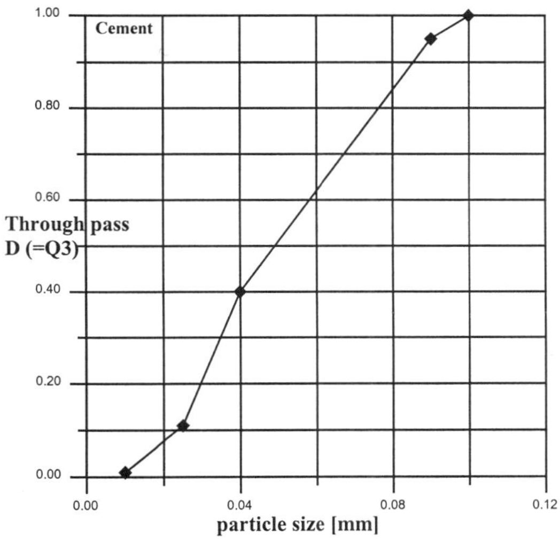

Abb. 4.3 Grain size of the cement - cumulative distribution (through pass) Q3

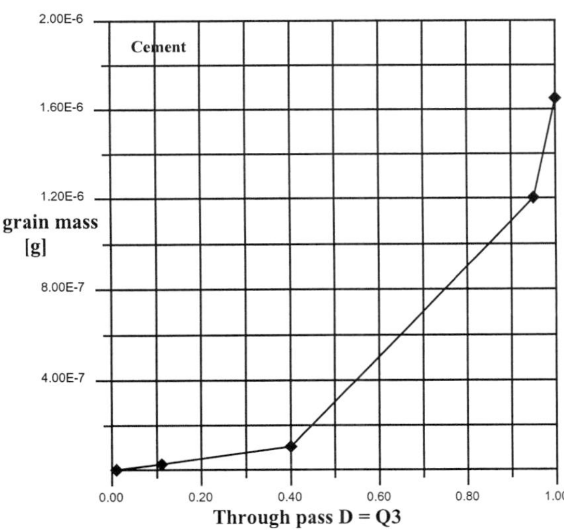

Fig. 4.4 The cement's grain mass in relation to the throughpass Q3

If you now integrate the grain volumes $\hat{m}_{p,i}$ and $\hat{m}_{q,i}$ as a function of through pass D you then get the required products2:

---

2 Note: For the dicrete integration the grain volumes in the middle of the intervals $\left(D+\frac{\Delta D}{2}\right)$ have been considered.

$$m_p\left(1+c_p^{\ 2}\right)=\int_{D=0}^{1} m_{p,i}(D)dD \ =0{,}121g$$
$$m_q\left(1+c_q^{\ 2}\right)=\int_{D=0}^{1} m_{q,i}(D)dD \ =4{,}51\cdot 10^{-7}g \tag{4.10}$$

The values of the figures in the process of calculations just presented and those in the diagrams are shown in Tables 4.2 and 4.3 so that the reader can check his own workings.

Table 4.2 Calculating the product in Stange's formula for sand

Grain size class	Mass proportion dQ=dD	Through pass D=Q3	Upper interval limit	Upper grain mass	Grain mass in
[mm]	[-]	[-]	dp,sand,i [mm]	[g]	Interval mean point ·dD
0-0.1	0	0	0.1	1.39E-06	0.00E+00
0.1-0.5	0.01	0.01	0.50	1.73E-04	8.64E-07
0.5-1.0	0.15	0.16	1	1.39E-03	1.17E-04
1.0-3.0	0.4	0.56	3	3.75E-02	7.77E-03
3.0-7.0	0.44	1	7	4.76E-01	1.13E-01
				Stange's product	**4.51E-07 g**

Table 4.3 Calculating the product in Stange's formula for cement

Grain size class	Mass proportion dQ=dD	Through pass D=Q3	Upper interval limit	Upper grain mass	Mass volume in
[mm]	[-]	[-]	dp,cement,i [mm]	[g]	Interval mean point ·dD
0-0.01	0.01	0.01	0.010	1.65E-09	8.25E-12
0.01-0.025	0.1	0.11	0.025	2.58E-08	1.37E-09
0.025-0.04	0.29	0.4	0.040	1.06E-07	1.90E-08
0.04-0.09	0.55	0.95	0.090	1.20E-06	3.60E-07
0.09-0.1	0.05	1	0.100	1.65E-06	7.13E-08
				Stange's product	**4.51E-07 g**

With a sample size of 2 kg you get a variance of:

$$\begin{aligned}\sigma^2 &= \frac{0{,}8\cdot 0{,}2}{2kg}\left[0{,}8\cdot 4.51\cdot 10^{-7}g+0{,}2\cdot 0{,}121g\right]\\ &= 1{,}94\cdot 10^{-6}\end{aligned} \tag{4.11}$$

for an ideal random mix.

Both ingredients, the sand and cement, differ plainly in their average particle size. This difference is made even more distinct when comparing their particle mass.

Consequently when calculating the ideal variance only the grain distribution of sand, the significantly coarser ingredient, is considered. Stange himself has consequently the simplified his equation as follows for this specific case:

$$\sigma^2 = \frac{pq^2}{M}\left[m_p\left(1+c_p{}^2\right)\right] \quad \text{für } m_p >> m_q \text{ ; } p \cong q \tag{4.12}$$

4.3 The Influence of Sample Size on the Variance of an Ideal Random Mix

The sample size M appears in the denominator and therefore the variance of an ideal random mix is inversely proportional to the sample size M:

$$\begin{aligned} &\frac{\sigma_1^2}{\sigma_2^2} = \frac{M_2}{M_1} \Rightarrow \sigma_2^2 = \frac{\sigma_1^2 \cdot M_1}{M_2} \\ &M_2 = 1\,\text{kg:} \quad \sigma_2^2 = 4{,}58 \cdot 10^{-6} \\ &M_2 = 10\,\text{kg:} \quad \sigma_2^2 = 4{,}58 \cdot 10^{-7} \end{aligned} \tag{4.13}$$

Thus the choice of sample size essentially determines the mix quality figure, although the mix's quality is the same, namely an ideal one, for all three sample sizes.
Upshot: you can only draw a comparison between mix quality figures where the same sample size is involved.

The variance is consequently not an absolute measurement of mix quality. Its numerical value changes with the sample size and also, for example, with the unit of concentration x or μ. Furthermore, you can only compare figures for mix quality if the average concentrations p or μ are the same.

5 Measuring Mix Quality Using Off and On Line Procedures

It was explained in the previous chapter how test values are analysed using statistical methods thus enabling a mixing process's efficiency to be judged. The influence of the size of the samples on an analysis, as also that of systematic or random sampling was discussed. The focal point of this chapter is that of how these mixing data are obtained. The methods adopted to do this can be divided into **off** and **on line procedures** (see Fig. 5.1).

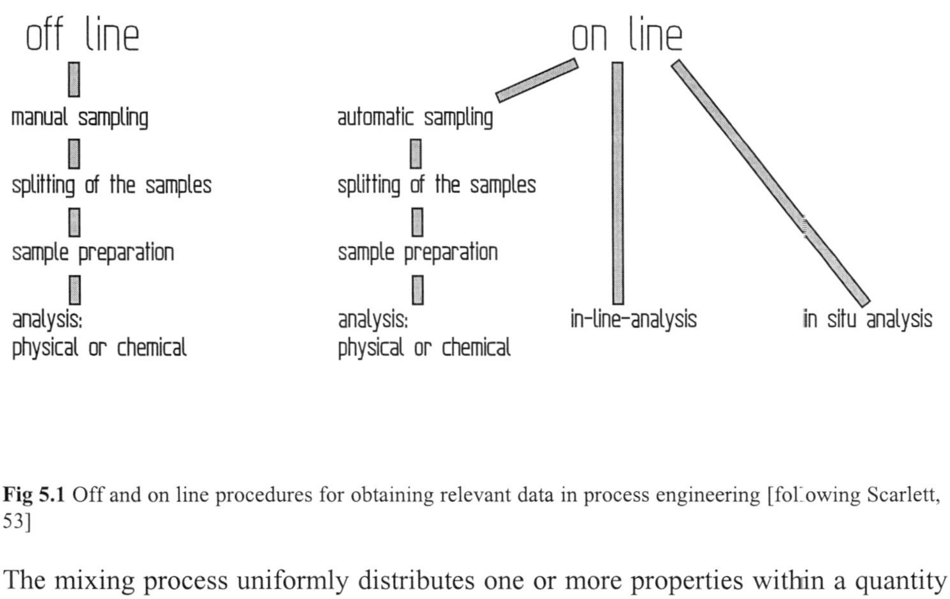

Fig 5.1 Off and on line procedures for obtaining relevant data in process engineering [following Scarlett, 53]

The mixing process uniformly distributes one or more properties within a quantity of material. These can be physically recordable properties such as e.g. size, shape, moisture or colour. Frequently however it is the mixing of **chemically** differing components which forms the subject under examination.
Off line procedures are shown as a flow chart in the left half of Fig 5.1. A specified portion is (randomly or systematically) taken from the volume of material. These samples are often too large for a subsequent analysis and must then be splitted. Many analytical processes, e.g. the chemical analysis of solids using infra red spectroscopy, require the samples to be prepared beforehand. At all of these stages

there exists the danger that the mix status within the samples will be changed. As a consequence, when examining a mixing process whose efficiency can be characterised by the variance expression $\sigma^2_{process}$, all off and on line procedures only give this variance only indirectly:

$$\sigma^2_{observed} = \sigma^2_{process} + \sigma^2_{measurement} \tag{5.1}$$

The observed variance $\sigma^2_{observed}$ also contains the variance $\sigma^2_{measurement}$ resulting from the test procedure and which arises out of errors in the systematic or random taking, splitting and preparation of the samples and from the actual analysis. A lot of attention is often paid to the accuracy of an analyser when it is being bought. However, the preceding steps of sampling and preparation also have to fulfil exacting requirements so that the following can apply:

$$\sigma^2_{process} >> \sigma^2_{measurement} \Rightarrow \sigma^2_{process} = \sigma^2_{observed} \tag{5.2}$$

Advances in sensor technology and data processing are enabling an increased number of procedures to be completely monitored using **on line procedures** (Fig. 5.1). Scarlett [53] counts **in line** and **in situ** procedures amongst on line ones. The great leap forward from off line to on line procedures lies in the fact that the whole process of preparing and analysing samples has been automated. As a result of this automation the amount of collectable test data has risen considerably, thereby enabling a more comprehensive statistical analysis and, in ideal cases, even regulation of the process. On line procedures must in most cases be precisely matched to the process and the expense in terms of equipment and investment is disparately higher. The accuracy of laboratory analyses in the case off line procedures cannot be produced using on line processes. Consequently on line procedures have to be regularly checked by (off line) tests in a laboratory. On line processes mostly use physical effects such as e.g. light scatter to determine particle sizes. There are as yet few on line procedures for chemically analysing solids. Near infrared spectrometers fitted with fibre optic sensors are used solely in the field of foodstuffs and for identifying raw materials in the pharmaceuticals industry and have also been applied to mixtures [54,61]. These record the specific adsorption of groups of chemicals on a particle's surface. If these spectrometers are based on modern diode array technology a spectrum covering the whole wave range is obtained in a fraction of a second.
In line and *in situ* procedures are also counted amongst on line ones. In situ procedures avoid the need for samples to be taken and the subsequent operations of preparing them (see Fig. 5.1). In line is taken to mean local measurements taken directly during the mixing process. Here local sensors should of course disturb the process as little as possible. In situ is taken to mean direct, integral measurement over a cross section without the sensor coming into contact with the material being processed.

5.1 Sampling Procedures

The purpose of taking samples is to record the properties of the whole volume of material from a small, analysed portion of it. This is difficult to achieve with solids since industrial mixes in particular always present a distribution of grain sizes, shape or density and can also separate out when samples are being taken on account of the ingredients' specific motional behaviour (cf Chapter 3).
Allen [55] formulates 2 *golden rules of sampling* which if applied avoid segregation when samples are being taken:

> Rule 1:
> Samples should be taken from solids whilst they are in motion.
>
> Rule 2:
> It is better to intercept the whole momentary flow of material several times at short intervals than to continuously draw off a portion from that flow.

Notes on rule 1: Where a quantity of material is at rest it is very probable that the coarser particles will concentrate near the surface (cf Chap. 3). For mixing technology this means that you do not take samples from inside the mixer but from the flow of material as the mixer is being emptied. This is all the more justified by the fact that the mixing process is not completed until the material has left the mixer chamber because the ingredients can also separate out when the mixer is being emptied as a result of the transloading operation and disturbance of the mix's strata.
Material should only be taken from the blending chamber in exceptional, warranted cases. Special sampling probes have been designed by Sommer [50], for example, to take samples of roughly the same size from the lower layers. Lacey [as discussed by Stalder, 56] has investigated the error sources of sampling probes.
Notes on rule 2: At each division of the flows of material and the selective examination of only one subflow there exists the risk that the substream under examination is not representative of the whole mix. This is illustrated by Stalder [56] who captured the discharge flow from a continuous mixer in a truss of cuvette trains and performed an image analysis upon it. Fig. 5.2 plainly shows that there is a clear difference in terms of both average concentration and mass flow between the individual trains of cuvettes. In this case taking the data from one train of cuvettes alone would certainly have given a false account of the mixing process.

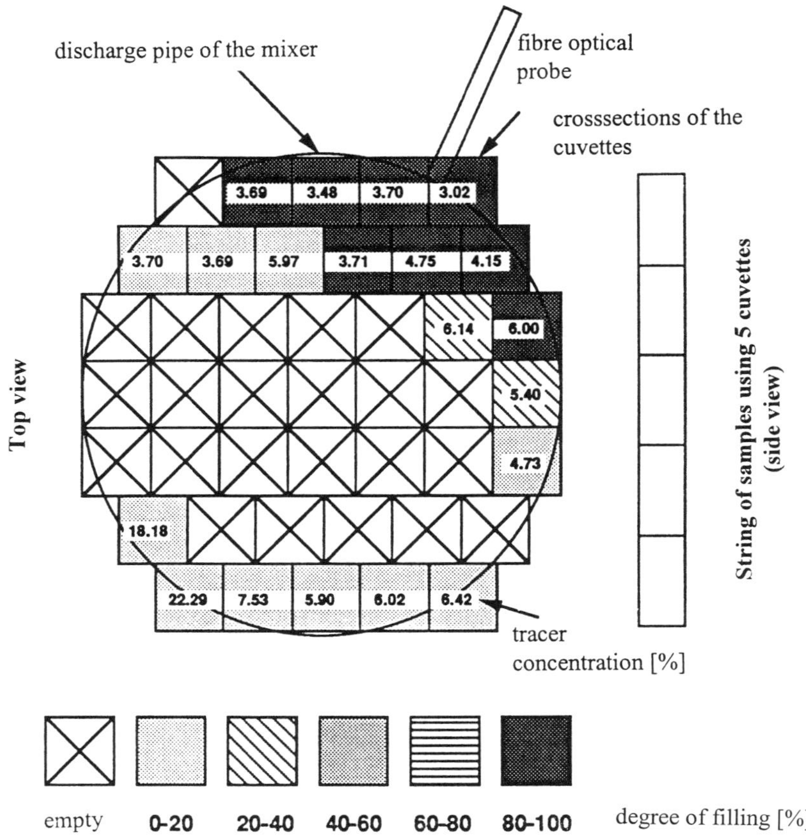

Fig. 5.2 Distribution of concentration and mass flow (degree of filling) across the cross section of a flow of material from a continuous mixer. The flow of material was trapped in a string of cuvettes and the SiC tracer component's concentration in the $Al(OH)_3$-SiC-mix was recorded using image analysis.

Analysing a solids mix is an expensive business and chemical analyses with the preparation of samples which is entailed are very expensive. That is one of the reasons why examinations of a mixing process are mostly carried out using suitable mix recipes allowing physics based methods being used to ascertain the mix quality. The following are widely used:

- electrical conductivity (e.g. salt/sugar mixtures) [68]
- examination of the differences in the ingredients' granular structure
- reflexion measurements in the case of ingredients of different colours
- radioactive identification of tracer particles [69, 10]

The results of random sampling can be processed with very little mathematical effort. Additional random sampling procedures have to be adopted in practice, such

as layer or multi-stage sampling which has to be additionally considered in the equation with regard to the cost involved. In his text book "Taking samples from powders and granular bulk products" K. Sommer [15] comprehensively postulates both the statistical principles and the relevant procedures involved, which is why we have not gone into them here.
Stalder [56] gives a good example of the analysis of powder mixes.

Photograph of the sample | Grey value measurement

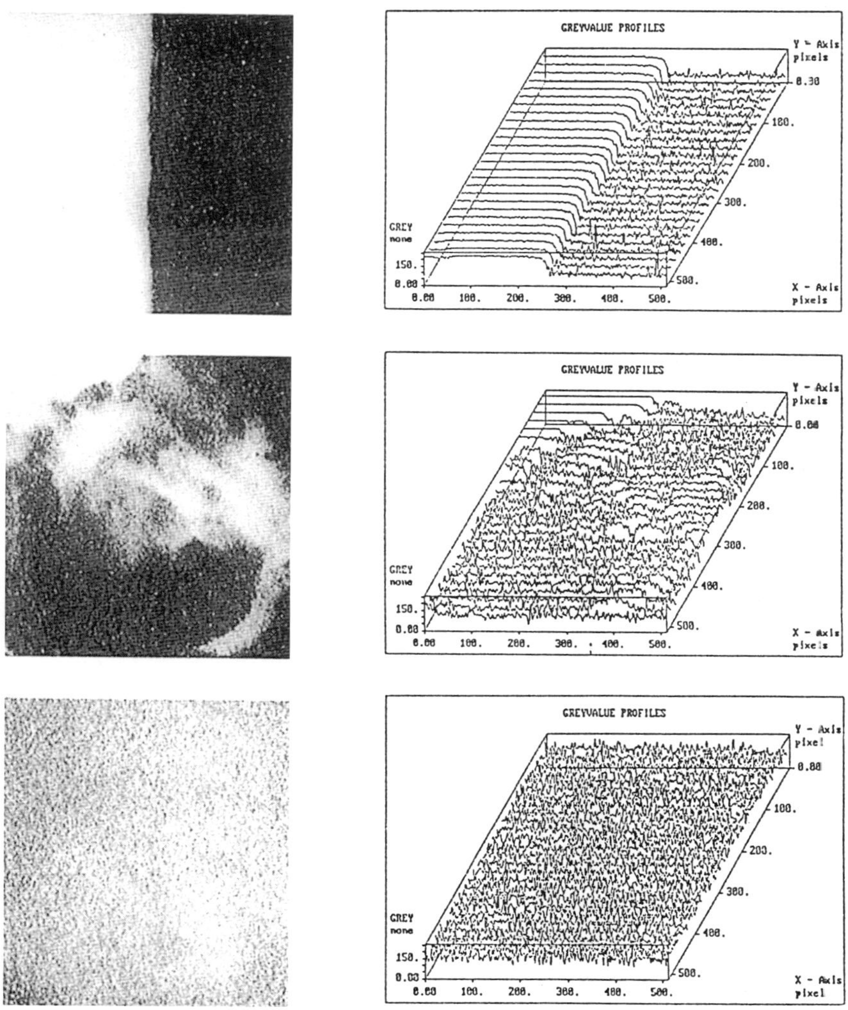

Fig. 5.3 Stalder's example [56] as recorded on video film of dark SiC being mixed with white $Al(OH)_3$ and an image analysis (grey value profile) of the same.

Fig. 5.3 shows three video frames recording the mixing of grey silicon carbide with white aluminium hydroxide. In the top frame the components are still separated. A microscope was used as the video recorder's lens and the video image was digitised using an image analyser. Differences in the powders' luminosity show up in a linear distribution of their grey values, with black corresponding to a grey value of zero and white to a value of 255. The grey value profile is shown on the right and from it the mix status can be seen in a two dimensional form. You can recognise the boundary between the ingredients by the sharp drop in grey value. Stalder mixed the ingredients with a spatula. The white stripes which can also be identified in the corresponding grey value profile can be clearly recognised in the second picture on the left. After thorough mixing a uniform mixture is achieved (bottom picture). Compared with the topmost picture the grey value distribution shows only minor fluctuations. Image analysis is also used by Dau et alii [70] for examining mixes.

5.2 A Fibre Optic In Line Method for Examining Mix Processes

An optical in line test system is presented in the following. In line, like in situ processes belong to the on line procedures (cf Fig. 5.1). For monitoring the process the sensor is placed directly in the mix and can register its local quality at that point. It goes without saying that the sensor should cause as little interference to the process as possible.

Development history

Mix models utilising the ingredients' differing optical properties have been employed since early on. Thus Oyama put forward a photographic method [57] in 1935 for monitoring the mixing process. The transparent, cylindrical end piece of a batch mixer was photographed at set intervals during the mixing process, for example after 3.8 and 4.9 revolutions of the mixer (Fig. 5.4). In the mixer there was a light and a dark powder. The film was developed and held in front of an adjustable slot. Light fell through this slot onto the developed film. According to the proportion of the darker ingredient, more or less light reached photoelectric sensors placed behind the film and which converted the light intensity into voltage signals. These activated a fibre potentiometer whose movements were recorded by a cinecamera. The smaller the fibre potentiometer's fluctuations per moment recorded the more homogenous the mix. Mix quality was defined in terms of intensity and not concentration and therefore included constants such as the period the subject was illuminated and the conditions under which the film was developed. Fig. 5.4 shows such an intensity curve. Oyama had already recognised that amplitude and frequency define the mixing process, i.e. the intensity and scale of segregation. However, an optical process for determining the condition of a mixture has the capability of being further developed into an in line, on line or in situ procedure. Improved detectors as well as fibre optic probes which allow the light source and detector to be spatially separated at the test point have become available in the meantime.

Kaye et alii [58], Harwood, Davies et alii [59] and Satoh et alii [60] use fibre optic probes to determine the concentration of a tracer in line. Whereas this method is proposed in the first of these works it is used extensively by Satoh to establish the properties of solids mixers.

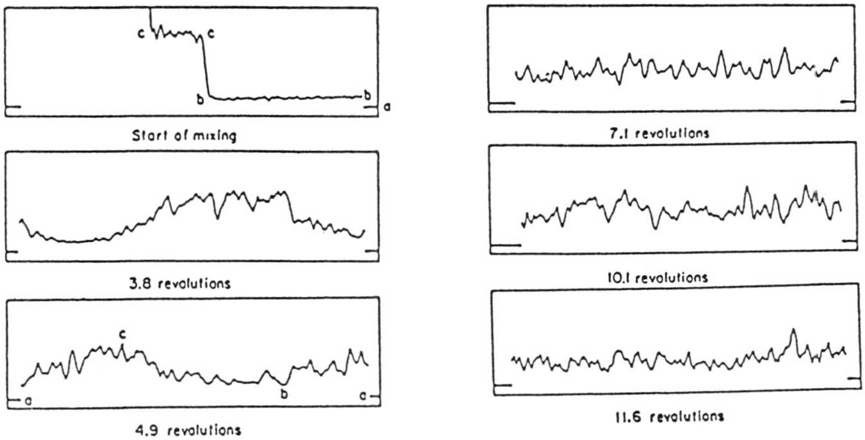

Fig. 5.4 Oyama, 1935: The first optical method for evaluating powder mixtures [57]. The deflection of a fibre potentiometer marks the proportion of a dark trace element in a mixture consisting of a light and a dark powder. The components are separated at the beginning of the investigation and as the mixing time increases, represented here by the mixer's number of revolutions, a mix is produced thus giving rise to smaller potentiometer fluctuations.

A fibre optic in line procedure was also used in the Federal Institute of Technology [ETH] Zurich to investigate continuous mixing processes [30]. The concentration of a tracer in the output from such a process was established in line using that method, thereby giving a comprehensive specification of a continuous mixing operation (cf Chapter 8). This in line test system's design and configuration is presented below.

The principle of in-line fibre optic measurement

A basic requirement of this method is that there should be a difference in colour between the two ingredients to be mixed since light will then be absorbed differently by them. Monochrome light is conducted to the powder mix by a light conducting fibre (Fig. 5.5). Because of their specific absorption capacities the two ingredients reflect this light differently. The proportion of the light diffusely reflected by the mix depends chiefly on the ratios of the two ingredients in the volume of material tested and upon the probes' geometry. If one of the ingredients is taken as a tracer there is a direct relationship between the tracer's concentration and the amount of the reflected light.

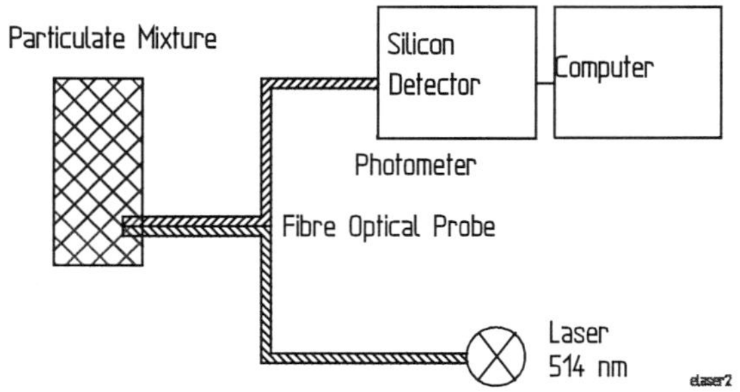

Fig. 5.5 Principle of the ETH-Zurich fibre optic measurement device. To calibrate the device the mix is stirred in a sealed container. The probe penetrates the side wall of the container through a hole drilled in it.

Light source and sensor

In most cases monochrome light is passed through the sensors onto the mix. The proportion reflected is measured with a highly sensitive photometer (Fig. 5.5). Using a white light source somewhat reduces the measuring system's sensitivity since the photometer is not just recording the reflection in a specific wavelength but across all of the white light's waveband. Instead of a simple sensor you can however use a spectrometer which records the specific reflection on each wavelength [61]. Spectrometers fitted with fibre optic light guides for the ultraviolet, visible and near infra red wavebands are available on the market. Measurements can be taken in a fraction of a second across the spectrum with particularly fast diode array spectrometers.

Three types of probes are shown in Figure 5.6, a coaxial probe bundle and two paired probes. The aperture angle and thus the emission and sensing cone is described by the numerical aperture (N.A.) which is defined as a function of the angle of refraction ϑ and the refraction index n_b in equation 5.3:

$$\text{N.A.} = n_b \cdot \sin\vartheta \tag{5.3}$$

The emission and sensing cone is indicated in Fig. 5.6 by lines of dashes. Reh and Li developed the angled probe (2.2) to determine local solids proportions in fluidized beds [63]. But this design is also an advantage in the case of straightforward solids mixes, an example being that with this type of probe the reflected light output is only slightly affected by the components' particle size [62]. Because of the angled configuration the emission and sensing cones penetrate just in front of the tip of the probe. As a consequence this probe has a large metering volume compared with others. However, this volume cannot be calculated exactly since reflected light from outside of the emission cone can be detected as a result of the manifold processes of reflection. This is certainly a disadvantage with regard to traditional sampling since using variance as a measurement of the quality of a solids mixture is only meaningful when the sample size is detailed at the same time (cf. Chapters 3 and 4). In the case of fibre optic probes the size of the sample and therefore the quantity of solid particles "seen" by the sensor can only be estimated [62]. The sample size for the solids mixes specified further down is about 0.01 grams for an angled probe, corresponding to a figure of 400,000 $Al(OH)_3$ particles. Since in line testing uses systematic measurements, the relevant values for larger sampling volumes can be estimated by means of equation 3.19 and taking the variance obtained with the probe.

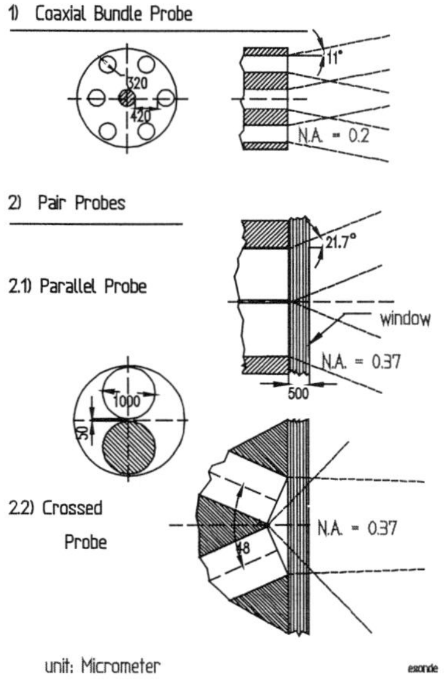

Fig. 5.6 The geometry of fibre optic probes for the in line recording of the quality of solids mixes where there is a difference in colour between the components (ETH- Zurich)

Solids configuration

Probes have been used with mixtures consisting of two ingredients which present a clear difference in their optical properties. White aluminium hydroxide $Al(OH)_3$ (Martinal 0-313, Martinswerk GmbH, Bergheim-Erft (Ger); A-L Alusuisse-Lonza-Gruppe) is added to dark silicon carbide SiC (100NN; Lonza Werke GmbH, Waldshut (Ger); A-L Alusuisse-Lonza-Gruppe) or a green pigment (Irgalite Green GLN; Ciba Geigy). **The dark components** are termed **the tracers** in the following. Table 5.1 gives an overview of the test results. Since the reflected light output is also influenced by the size of particles in the ingredient, it is only possible with this in line system to make precise statements if the ingredients have a narrow grain size distribution.

Table 5.1: Test materials

	$d_{p3,10\%}$ [μm]	$d_{p3,50\%}$ [μm]	$d_{p3,90\%}$ [μm]	ρ_S [g/cm³]	ρ_{bulk} [g/cm³]	colour
$Al(OH)_3$	2	27	72	2.43	1.31	white
SiC	7	24	56	3.22	1.51	black
Irgalite	0.05	0.22	2.3	2.13	0.33	green

Measurement range

Fig. 5.7 shows the measurement range of the fibre optic in line method used on a mixture of SiC and $Al(OH)_3$ [62]. Here the averaged reflection signal R is projected across the concentration of the dark tracer component SiC in the SiC-$Al(OH)_3$ mixture. The reflected light output is related to the signal produced by the pure white aluminium hydroxide under the same conditions:

$$R = \frac{\text{signal}}{\text{signal}_{0\%\ \text{tracer}}} \tag{5.4}$$

Using an angled paired probe enables concentrations of down to 7% SiC in the mix to be measured under this composition of particles. The corresponding values for an Irgalite/$Al(OH)_3$ mix can be found in [30, 62].

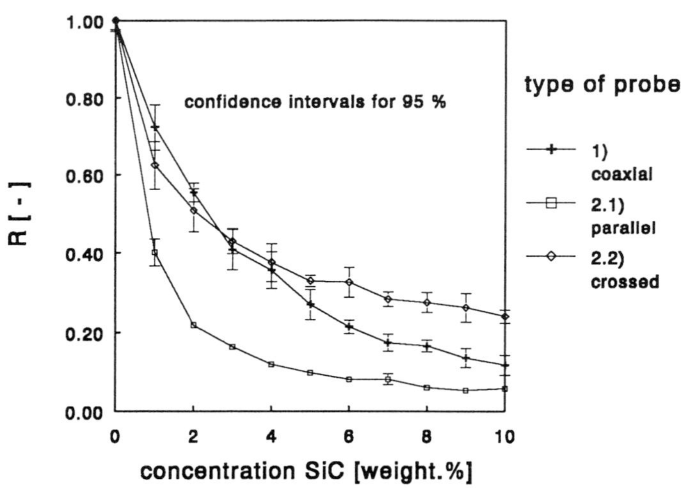

Fig. 5.7 Averaged signal of the back scatter light output for an angled probe and an SiC/$Al(OH)_3$ mix [62]

In line concentration measurement using a fibre optic probe in the flow from a continuous mixer

As soon as the relationship between an ingredient's reflected light and concentration is known (Fig. 5.7) this measurement system can be employed to define mixing processes in line (Fig. 5.8).
The probe is mounted at the outlet of the continuous solids mixers. An adjustable gap assures a permanent covering of the probe (Fig.5.8). A packing of particles of some 3 cm height is built up in front of and above the probe, continuously being replaced by the mixer discharge flow. In intervals of 0.2 seconds the intensity of the reflected light is measured and the variation in time of tracer concentration computed therefrom.

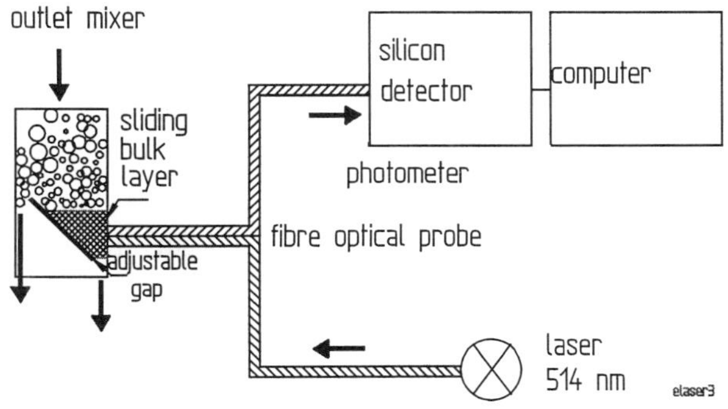

Fig. 5.8 In line measurement of the concentration at the mixer's outflow

Fig. 5.9 shows a stimulus-response-experiment. 2000 signals were recorded during the period of measurement. The mixer ran at a steady state and both ingredients were continuously fed into it. After 80 seconds the SiC tracer flow was cut off. The compound then showed a noticeable reduction in SiC until finally only white $Al(OH)_3$ came out of the mixer. As expected the diffusely reflected light output increased markedly after the dark tracer was turned off. A concentration curve can be calculated from this signal pattern if the functional relationship between light output and the tracer's concentration has been established from calibration tests (see Fig. 5.7).
Fig. 5.9 shows the concentration curve. In steady state operation the SiC concentration hovered around 1.2%. After the SiC tracer was turned off the concentration fell back to zero. The residence time distribution in a continuous mixer can be determined on the basis of such trials and we will go into this in greater detail in Chapter 8.

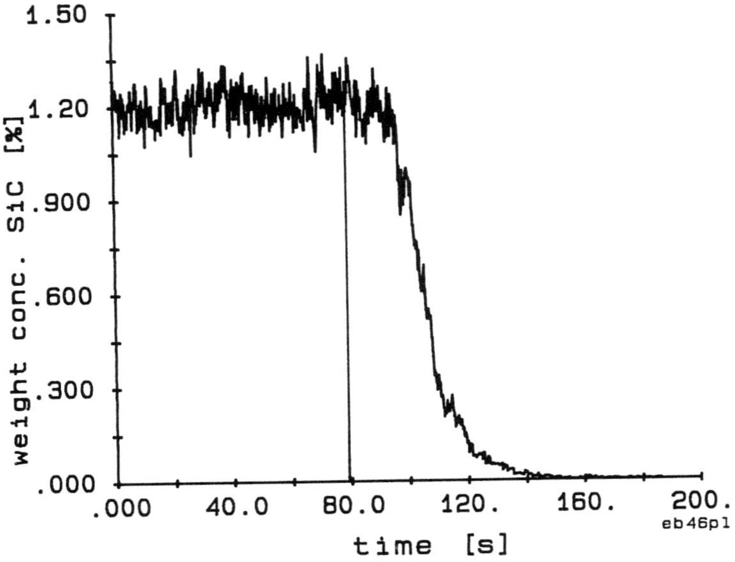

Fig. 5.9 Stimulus - response experiment: variation in time of tracer concentration, SiC-Al(OH)$_3$ - mixture, SiC-feed stopped after 80 seconds, average SiC concentration before step = 1.2%

In industrial practice you seldom in fact use mixes whose mix condition can be directly recorded by the procedures we have just presented. However, with this in line process and the appropriate mix recipes it is possible to obtain universally applicable data regarding the process of mixing and to design and improve solids mixers.

6 Equipment for Mixing of Solids

A huge variety of equipment is available on the market to suit the multiplicity of mixing tasks. Ries [66] gives a detailed overview and his classification of mixers is partly adopted in the book written by Wilke et alii [67]. Müller's general article [26] represents an excellent overall synopsis. He draws a distinction between large capacity bunker blenders in which only a part of the material is moved and those which produce a mix by moving the entire mixer or by rotating agitators fixed inside it. Besides these units there are special mixed stockpiles for homogenising large volumes of materials. A mix can also be produced by concentrating locally the controlled feed of the components.

In this chapter mixers and devices for mixing solids are divided into four groups as follows:

- mixed stockpiles
- bunker mixers
- rotating mixers or mixers with rotating tools
- mixing in the feed flow

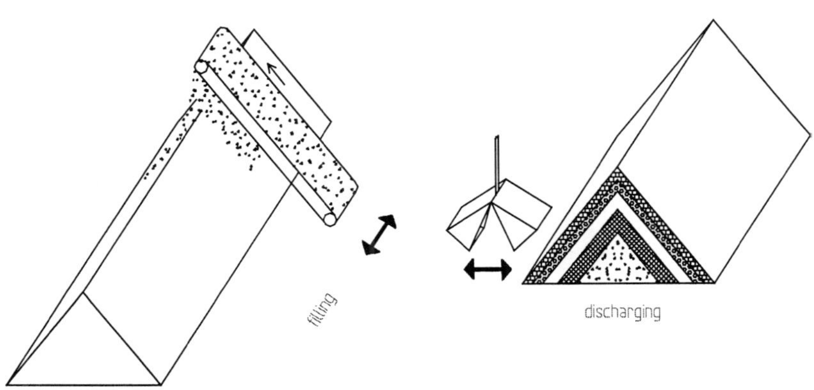

Fig. 6.1 Long stockpiles for homogenising large volumes of material

6.1 Mixed Stockpiles

Many bulk goods which are often stored in very large stockpiles do not possess uniform material properties within these stockpiles. In the case of raw materials this is caused for example by natural variations in deposits and in the case of primary material by variations between different production batches, e.g. in the iron and steel industry there are fluctuations in the ore and carbon content of the finished material. If these stockpiles are emptied in the "first in first out" principle, material with a variance in its properties will find its way into the subsequent process and reduce its efficiency.

Fig. 6.2 Charging coal into a homogenising stockyard (Central Coking Plant Saar GmbH, Germany) capacity of the stacker: 1200 t/h; boom length: 47.5 m (Courtesy of PWH – Krupp Engineering)

To provide a uniform finished material a mix is obtained by following a defined scheme for building up and emptying large stockpiles. Such mixing processes are also called *homogenisation*. Fig. 6.1 shows two longitudinal stockpiles. The first illustration shows the typical creation of a long stockpile whilst the other shows its discharge. As in any mixing process the volume of material is homogenised by moving portions of it relative to each other. A long stockpile is built up by a movable conveyor belt or other corresponding device travelling *lengthwise* (Fig. 6.2). During loading the belt continuously travels up and down the whole length. In the strata thereby created is stored a temporal record of the material's delivery. If the material is now systematically removed *crosswise* to these layers each portion removed from the stockpile (Fig. 6.3) will contain material from all the strata and therefore from the times it was supplied. Since such bins are built up over days or weeks, mixed stockpiles reduce the degree of long term fluctuations in the material's properties.

Fig. 6.3 Recovery of the fine homogenised coal by system Chevron (Central Coking Plant Saar GmbH, Germany); width of the bridge scraper: 57.5 m ; capacity 1200 t/h; (Courtesy of PWH – Krupp Engineering)

6.2 Bunker and Silo Mixers

Bunker and silo mixers are sealed vessels, the biggest of which likewise serve to homogenise large quantities of solids. They are operated batchwise, continuously or with partial recirculation of the mixture. Their sealed construction also enables material to be conditioned, e.g. humidified, granulated, dried or rendered inert, as well as mixed. Fig. 6.4 shows W. Müller's classification of bunker mixers [26].

In *gravity mixers* granular material is simultaneously drawn off by a system of tubes at various heights and radial locations, brought together and mixed. Other types of construction use a central take-off tube into which the solids travel through openings arranged at various heights up this pipe. If the quality of the mix does not meet requirements the material drawn off is fed back into the bunker. In this fashion the bunker's entire contents are recirculated several times and thus homogenised. The material drawn off is in most cases carried to the top of the bunker by air pressure (using an external circulation system). Gravity mixers are designed for free flowing powders and are offered in sizes ranging between 5 and 200 m^3. The specific energy consumption, i.e. the energy input per product mass, is very low at under 1 to 3 kWh/t. Silos with a special funnel mixer at their outlet are also counted in with the gravity mixer group (silo screw mixer). A concentric double cone gives a different residence time period for the material in the inner and outer cones, producing a remixing. This type is offered for quantities of material between 3 and 100 m^3.

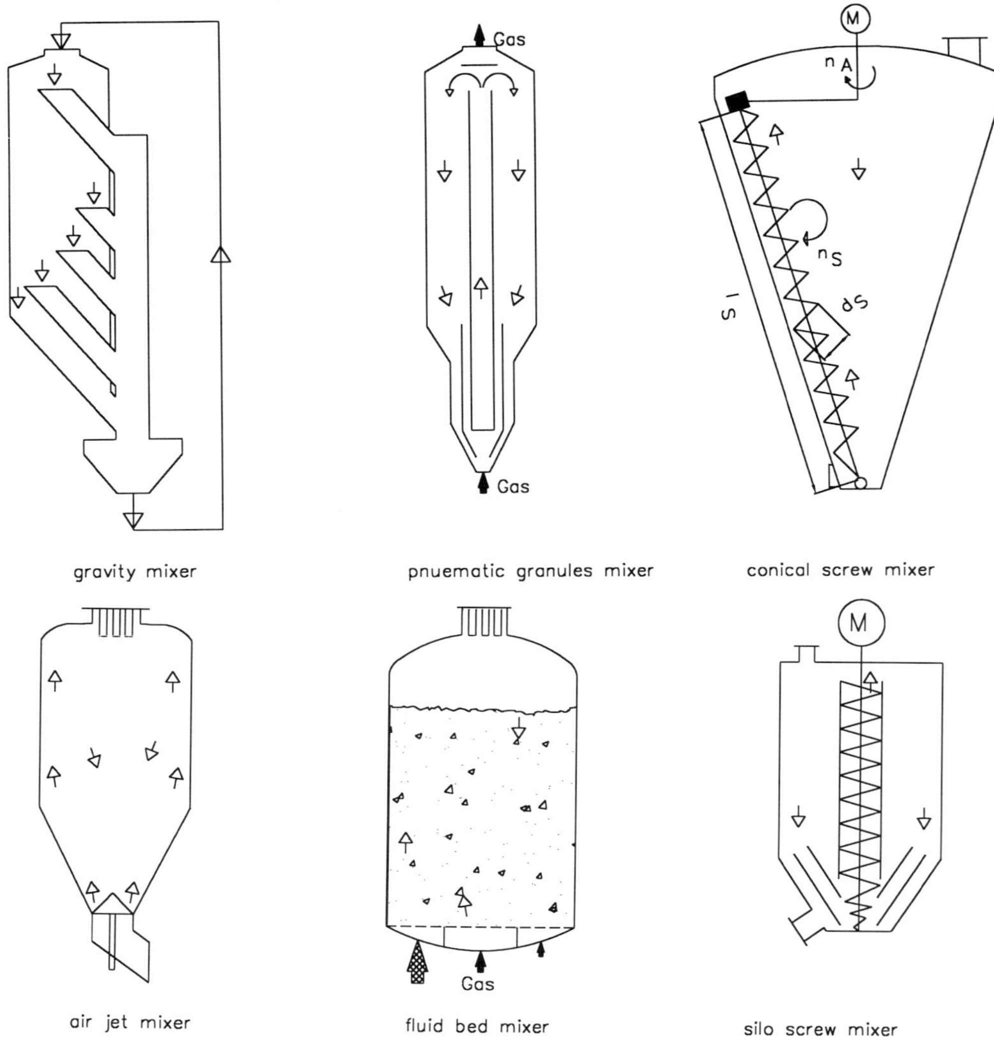

Fig. 6.4 W. Müller's classification of bunker or silo mixers [26]

Wilms [104] showed that the performance of such blenders is drastically influenced by the operation mode, batch versus continuous operation (Fig. 6.5). Most of these mixers which are mainly used in the plastic industry are operated batchwise (a). A tracer representing around 15 % of the total mass in the mixer is being brought onto the mayor component in the silo. Now the silo is being emptied and the concentration of the tracer component at the outlet is determined. During this trial

no mass is being recirculated (single pass). The batch operating blender distributes the tracer during the discharge but still the tracer concentration reaches values of 40%. I f the same mixer is operated in continuous mode meaning there is a constant flow of material $\dot{m}$ fed into the mixer and being discharged (b) the tracer reaches values of less than 20 %.
Eichler and Dau [105] measured and calculated the residence time distribution in such mixers (type Binsert). Fig. 6. 6 shows 2 peak of the differential residence times distribution showing clearly the two regions in the mixer with its own flow behaviour (the fast flowing material in the central region versus the slow flowing material in concentric wall region).
In *granulate mixers* material from various areas of the vessel is brought together in its lower section and then carried upwards by air pressure in a central pipe (using an internal circulating system) where the solids are separated from the gas and at the same time distributed on the surface (cf. Chapter 2 on "Mixing by division and combination"). Design sizes reach up to 600 m^3 and the specific energy input, like that of gravity mixers, is similarly low.

The rotating screw of a *conical screw mixer* transports the material upwards from the bottom. This screw is at the same time driven along the wall of the vessel by a swivelling arm. This type of mixer processes also both pastes and cohesive powders. The solids at the container wall are continuously replaced by the action of the screw so that the mix can be indirectly heated or cooled through the container's outer wall. It is also used for granulation and drying. Mixers of this design are offered in capacities of between 25 litres and 60 m^3. A variant of this mixer type drives the screw in a epicyclic path thus covering the whole space of the conical mixing vessel.
In *blast air* or air jet mixers air is blown in through jets arranged around the circumference of a mixing head placed in the bottom of the vessel. The specific air consumption is 10-30 Nm^3/t and the largest mixers have a capacity of 100 m^3.
If a fluid flows through a layer of particles against the force of gravity the layer loosens up when the fluid reaches a certain speed (the velocity of minimal fluidisation and the particles are held in suspension by it. As the fluid's velocity increases, the layer of particles kept in suspension by the gas flow extents although the pressure drop remains. The particles in the *fluidized bed* become ever more mobile [34]. Fluidized beds thereby possess excellent mix properties for solids in both a vertical and radial axis. In circulating fluidized beds in particular this is combined with elevated heat and material exchange values as a result of the high relative velocities of the gas and solids[1]. Lower fluidizing speeds to limit air consumption are used if the fluidized bed only serves the purpose of mixing. Furthermore, differing volumes of air are fed to the air permeable segments installed in the container's floor. The largest fluidized bed mixers as used in cement making reach a capacity of $10^4 m^3$.

1 A more extensive description of gas/solids reactors and their mixing properties for gases and solids is given by Reh [Chemie-Ingenieurtechnik 49 (1977), No. 16 P. 786-795)

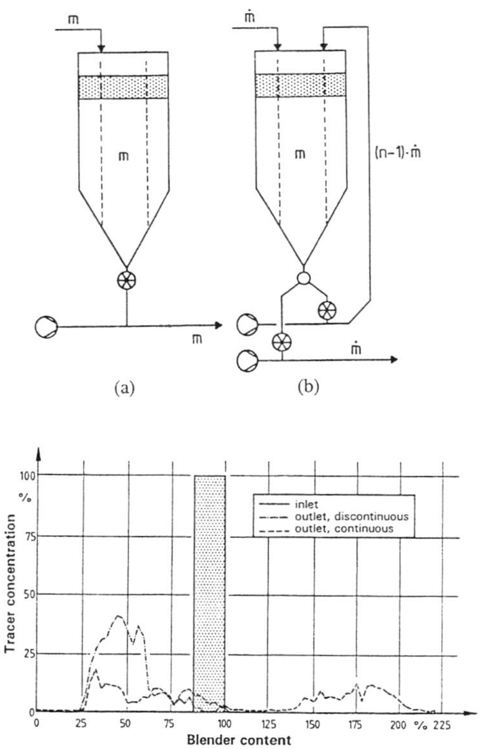

Fig. 6.5 Comparison of batch and continuous blending in silo mixers (single pass) following Wilms [104] Influence of the operation mode on the mixer performance: a) batchwise b) continuous
The silos is filled with the main component. A layer of tracer particles is being filled onto the other component (inlet 85-100 % of content, 100 % concentration) . During discharge the tracer concentration is being determined at the outlet. The diagram of blender content („residence time") versus tracer concentrations shows that , for continuous operation, the amplitude of the tracer reaches only about 50 % of the value for batch operation, i.e. much better mixing

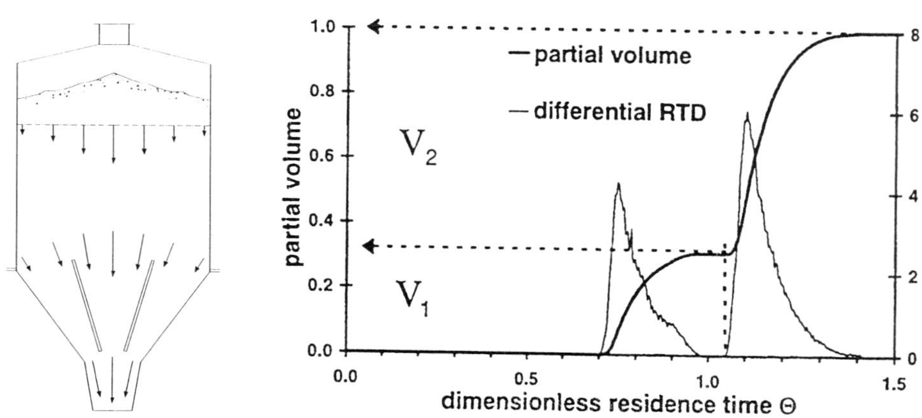

Fig 6.6 Residence time distribution in Binsert -mixers following Eichler and Dau [105]
The difference in velocity between the material in the center core of the mixer and the material in the wall region results in a differential residence time distribution (RTD) which shows 2 peaks
hopper half angle 45°, Binsert half angle 20 °, total volume 0.33 m^3

The material must be fluidizable, i.e. free flowing (with a particle size greater than 50 μm) and dry. The specific power input lies between 1 and 2 kWh/t but air consumption rises sharply in the case of particle sizes above 500μm. Fluidized bed granulators use the mixing properties of fluidization for granulation and here the suspension is atomized onto the fluidized bed.

6.3 Rotating Mixers or Mixers with Rotating Components

Fig. 6.7 shows 4 categories of mixers where the mix is agitated by rotating the whole unit or where movement in the mix is produced by rotating components built into the apparatus. Rumpf [26] classified the types of mixer according to their Froude number (Fr):

$$\mathrm{Fr} = \frac{r\omega^2}{g} = \frac{rn^2 4\pi^2}{g} \tag{6.1}$$

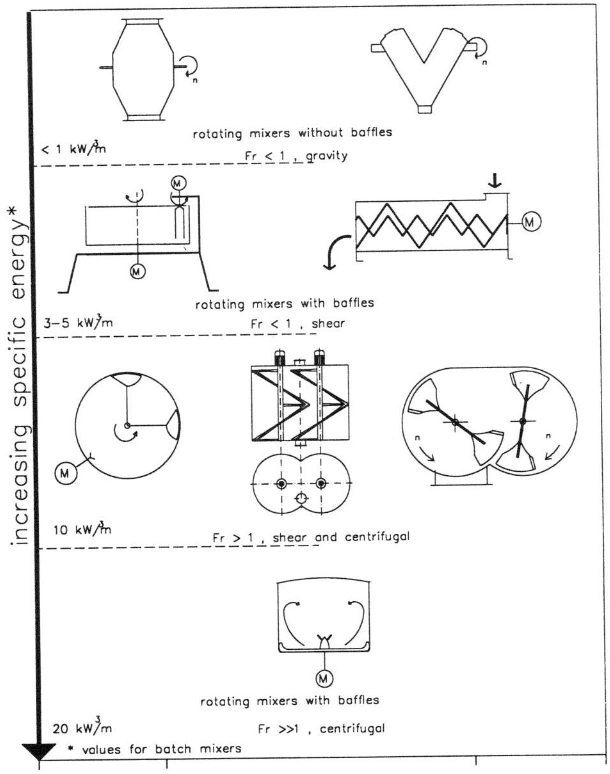

Fig. 6.7 W. Müller's classification of the movement of material by rotating agitators or revolving containers
* Specific output data applies to batch mixers. The *energy* applied to the mix is determined by the

product of the power and mix time.

Here r denotes the mixer's radius or that of the mixer's agitators, g the gravitational acceleration and ω the angular velocity. The Froude number therefore represents a dimensionless rotating frequency. The Froude number is the relationship between centrifugal and gravitational acceleration. No material properties are accounted for in the Froude number. Müller [26] points out that as opposed to the case of fluid mixing it has not yet been possible to define an indicator for solids similar to the Reynolds number. There is no meaningful relationship with a material parameter which might be comparable to viscosity. Thus the Froude number is only designed for comparing the different operating conditions of a type of mixer and does not describe any flow condition [30].

Fig. 6.7 shows four classes of solids mixers. Subject to the above limitations a distinction is drawn here between Fr <1, Fr >1 and Fr >>1. Müller puts *free fall mixers* into the first category, where the container is moved and thus the material is mixed inside it. Free fall mixers are only suitable for free flowing solids. A familiar example of a free fall unit is the drum mixer where in this model the mixing can also be overlaid with a segregating operation leading to complete separation of the ingredients. Since drums are also used in additional processes such as rotary tubular kilns or as granulating drums for solids, with these machines one has to watch for separation, e.g. in respect of particle size. However, mixers without built in agitators are particularly widely used in the pharmaceuticals and foodstuffs industries since they can be cleaned very thoroughly. *Asymmetrically moved mixers* in which for example a cylinder is tilted obliquely to the main axis, turning over the mix, also belong in the free fall category. Mixing is done gently. Because of the material's distance from the axle (see Fig. 6.7) high torques have to be applied by the drive motor and these moments have to be supported by the mixer's bearings and bed. Units with a capacity of 5000 litres are offered [67].

There are also mixers operating in the sub 1 Froude number range and where the work of moving the mix is undertaken by rotating agitators. The particles of solids are displaced relative to each other by agitators inside the mixer. This model is suitable for both cohesive, moist products and those which are free flowing. Examples of displacement mixers are ribbon blenders or paddle mixers. Because of their low rpm the load on the machine is slight but the mixing process is relatively slow. The specific energy input is low and lies under 5 kW/m^3.

Ploughshare, centrifugal and multi-flux fluid mixers operate in a Froude number range greater than 1. The consequence is that at least in the vicinity of the outer edge of the agitator the centrifugal forces exceed that of gravity and the particles are spun off. Thus instead of a pushing motion there is a flying one. This accelerates the mixing process both radially and axially. If the ingredients still need to be disagglomerated, high speed cutters are brought into the mixing space to disagglomerate the mix by impact.

Fig. 6.8 illustrates the mixing process in a Multiflux-mixer which is a double shaft mixer with overlapping mixing tools. This mixer usually operates just above the sequence. The mixer is filled with a bulk ingredients and on top, a black powder is placed. The mixer is started and the sequence shows how the material gets mixed. Mixing times of such mixer are very short. Depending on the size of the mixer they vary between 20 s to a few minutes [109].

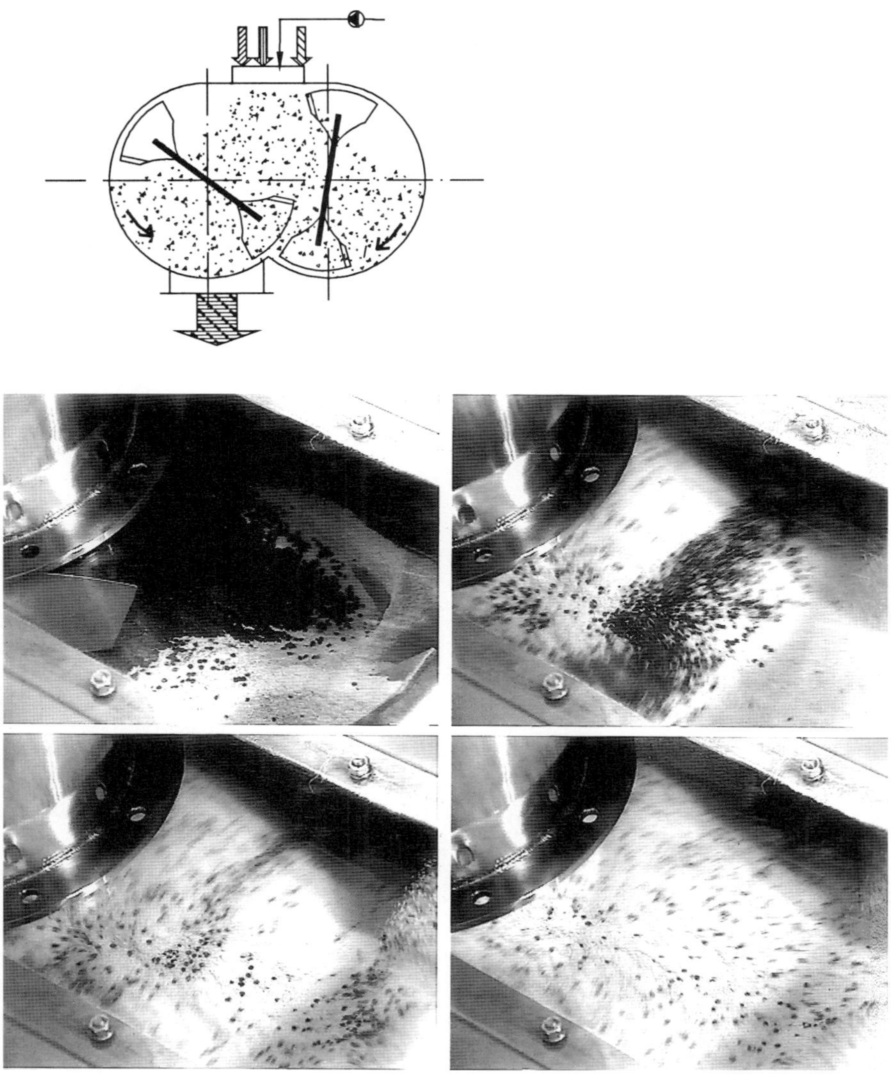

Fig 6.8: Batch mixer type Multi-Flux (Gericke): 2 counterrotating, synchronized and overlapping mixing tools which usually operate just above the critical speed, volume betweeen 25-5000 l
The videosequence illustrates over a time sequence of 20 seconds the mixing of black and white PE-Particles

At very high Froude number ranges (a Froude number greater than 7) there is a sharp increase in the shear forces acting on the mix. The impact load is enormous and the product heats up as a result of dissipated energy. The heat is caused by friction between the mixer's tools and the solids as well as by friction amongst the solids' particles. As well as simple mixing, here the mixer's task is often disagglomeration, agglomeration, moistening and sintering. Such mixers are especially used for producing plastics and in pharmaceutical industry for granulation.

6.4 Mixing by Feeding

Mixing in the feeder stream represents a continuous mixing process. The solids are blended by metering in each ingredient and bringing these streams of solids together locally. There is no axial mixing (transverse or back mixing) or as such it is very low, with the result that the quality of the metering determines the mix's homogeneity. Metered (gravimetric) feeder units should therefore ideally be used (see Chap. 9).

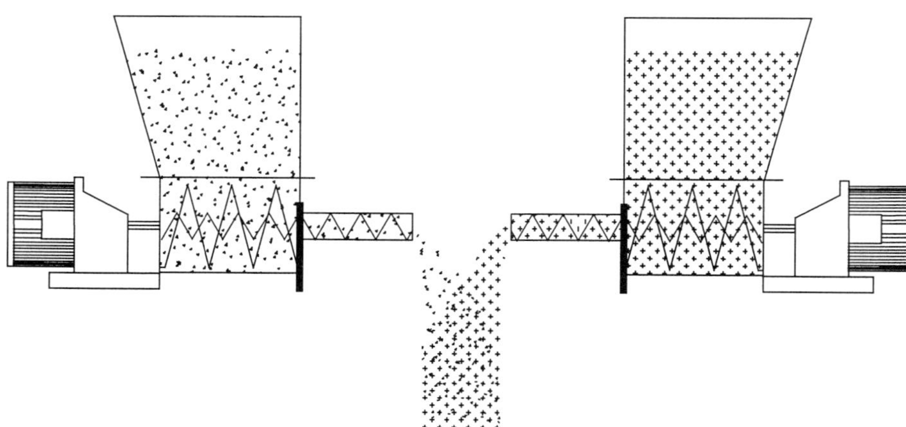

Fig. 6.9 Mixing in the feeder stream

According to the requirements of the case in question, mixing is also required obliquely to the direction of travel. If the ingredients are brought together in a perpendicular fall this is achieved by their merging together (Fig. 6.9). If this oblique mixing is not sufficient,*static mixers* can be used for free flowing powders or granules where, for example, the stream of solids is repeatedly divided up and brought back together by baffles as it drops down a tube. The energy input into the mixer is very low but such systems need height. Such mixers have been investigated by L. T. Fan [106] et alii and Gyenis et alii [107]. Eichstädt [108] pointed out that the efficiency for radial mixing depends on the gas phase as well. At best they operate with low volume concentration and for particles in the size range between 20 and 200 microns. Fig. 6.10 shows a static mixer which has been investigated by Eichstädt [108].

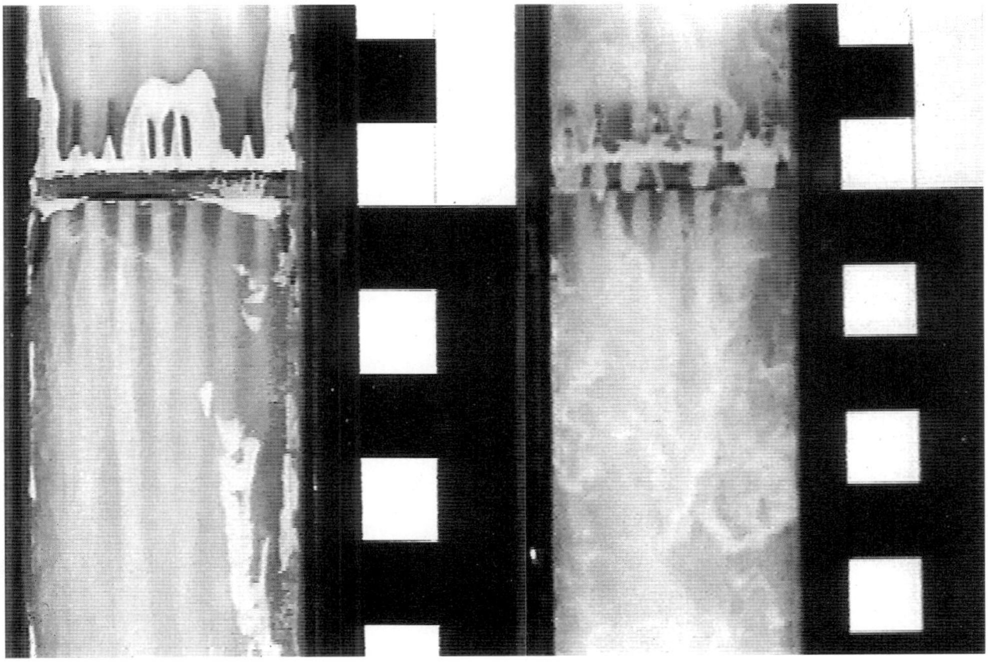

Fig. 6.10 Flow regime and radial mixing in a static mixer following Eichstädt [108]
The cylindrical mixer is continuously fed with $Al(OH)_3$ -particles [flow: 60 g/s), $d_{50,3}$= 30 micron], the particles are passing the mixer vertically. Obstacles in the flow create a mixing effect. If the entrance region of the static mixer is designed such that no ambient gas may enter with the solids, the radial mixing improves drastically
left: static mixer, gas can enter the mixer **right:** gas suction into the mixer is blocked

This Chapter has tried to give the reader an overview of solids mixers and their systematics. This is to help him in choosing a suitable type of unit for his own problems in the field mixing. However, each general illustration cannot list all the types of mixers and we would therefore refer him to the relevant manufacturers data.

7 Batch Mixing

Batch or *discontinuous mixing* is characterised by the fact that the mixer is filled with the ingredients, then started and after a certain time („mixing time“) the mixture is discharged. The feeding (filling), mixing and discharging operations are performed *one after the other*.
In the next chapter there follows a detailed comparison between continuous and batch mixing processes. Batch processing presents advantages for small quantities of material because of its lower investment costs and greater flexibility. Batch mixers are used even when very large volumes of material are being homogenised since continuous mixers are limited by their lower volume when it comes to backmixing.
However, in the batch mixer's very flexibility there lies the danger that it is not being optimally used and overmixing can for example occur, whereby the product could possibly be damaged and the process's effectiveness suffers. The time required by a mixer to mix ingredients which are initially separated basically determines its economic efficiency.

7.1 Determining the Mixing Time by Means of Sampling

It is not yet possible to theoretically forecast mixing times for solids and therefore these have to be ascertained by dint of experiments. The traditional method of determining mixing times is once again sampling followed by off line analysis. The mixer is loaded and started.
After the mixer has been loaded with the ingredients in accordance with a defined procedure it is run and samples are taken from it at set time intervals (cf. Chap. 3). To do this the mixer usually has to be halted. The concentration of the tracer in the samples is established and the random sample variance ascertained. This random sample variance serves as an estimated value for the variance which defines the mixture's condition (cf. Chaps. 3 and 4). All analyses are burdened by errors and

this is expressed in a variance derived from the sampling itself and from the analysis procedure (cf. Chap. 5). Sommer points [22] out that a recording of the variance's time curve is greatly influenced by the accuracy of the analysis. This is illustrated in Fig. 7.1 .

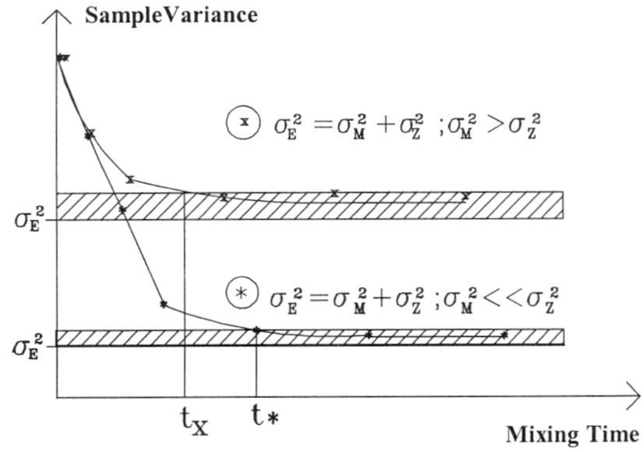

Fig. 7.1 Sommer's [22] illustration of the influence of the measurements' accuracy on the variance as a function of the mixing time, a set of samples have been taken at different mixing times for computing the sample variance. With increasing mixing time the sample variance decreases, the mixing time is the time for which no further improvement of homogeneity with longer mixing is achieved. But special attention has to be paid whether the experimental sample variance monitors the errors of the analysis procedure (x) or detects really the mixing process (*)

Confidence intervals for the final status σ^2_E are shown as hatched section.

x- the final status's variance σ^2_E is determined by the analysis procedure's variance that leads to erroneous setting of the mixing time t_x.

* - the final status's variance σ^2_E is determined by the mixing process's variance. The measured mixing time t* characterises the mixing process.

To begin with there is a sharp fall in the random sample variance and it runs asymptotically towards a final value of σ^2_E as the mixing time increases. This stationary end value is set by the variance of the mix in the stationary condition σ^2_Z, ideally the variance of an ideal random mix, and the variance σ^2_M caused by errors in the analysing process. The **mixing time** denotes that period in which the experimental random sample variances fall within the confidence interval of the stationary final condition. Two cases represented by "x" and "*" are illustrated in Fig. 7.3. σ^2_E in case "x" is determined by the analysing process itself. The mixing process's fluctuations in its stationary condition are much smaller than those arising out of the analysis. Therefore the mixing process can only be tracked at its commencement. The "mixing time" t_X obtained under these conditions does not characterise the process. In case "*" the analysing process is sufficiently accurate for the mixing process to be followed through to its stationary condition. Sommer [22] points out that the "mixing time" t_X obtained on the basis of an unsatisfactory analysis is always deceptively shorter than the actual time t_*.

7.2 Obtaining the Mixing Time Curve Using Fibre Optic In-line Measurements

Mixing times can also be determined with the probe technology illustrated in Chapter 5. There a fibre optic probe was used in the discharge from a continuous mixer. In a totally similar way such a probe can be mounted in the outlet of a batch mixer and thereby enable the mix quality obtained after the mixer is emptied to be defined. However, to ascertain the mixing process's time curve the probe has to be placed in the mixing area itself (Fig 7.2). To that end the probe in a horizontal mixer (Fig. 7.2, a model MXC-150 from Gericke) was fed into the mixing area through a bored hole, with the probe's window fitting flush with the mixer wall (Fig 7.3). The mixer was filled for the trials with the already familiar test mixture of silicon carbide (SiC) and aluminium hydroxide ($Al(OH)_3$).

Fig. 7.2 Horizontal mixer type MXC-150 (Gericke) with mixing tools shaped like plough shares used for experiments to determine mixing times, volume 150 liter, rotational speed 160 rpm

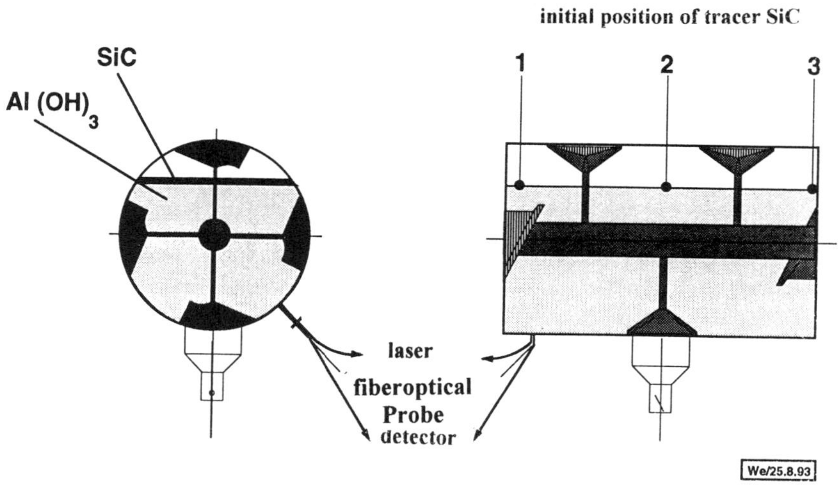

Fig. 7.3 Establishing mixing times in a MXC-150 (Gericke) horizontal mixer with a fibre optical probe. Parameters: the initial position of the tracer relative to the probe.

The tracer was added in the form of a narrow stripe on top of the $Al(OH)_3$. During the 3 experiments the axial location of the tracer relative to the probe was varied. In position 1 the dark tracer is close to the mixer's left hand end wall and thus facing the probe. In position 2 it is in the middle and in 3 the tracer is deposited on the aluminium hydroxide near the right hand end wall. Fig. 7.4 shows how for the relevant positions, i.e. that of the tracer relative to the probe, the concentration in front of it asymptotically approaches the tracer's average concentration in the mixer as the period of time stretches beyond point zero when the machine was started. At position 1 the probe abruptly "sees" very high concentrations of the tracer which are reduced as the mixing time increases. In plough share mixers axial is slower than radial mixing since mixing in the horizontal plane has to occur vertically to the plough shares' plane of rotation.

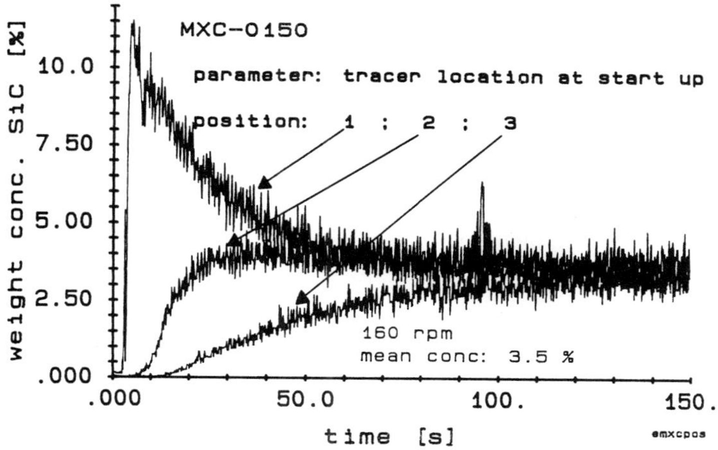

Fig. 7.4 Curves of mixing times in an MXC-150 (Gericke) plough share mixer obtained using fibre optic in-line measurement. Tracer: SiC (black), main ingredient $Al(OH)_3$ (white).
1) With the tracer at the left hand end wall near the probe. 2) With the tracer in the middle 3) With the tracer at the right hand end wall.

As already indicated above, the initial conditions basically determine the mixing time. In the worst case, with the tracer located at an end wall (positions 1 and 3), blending has to take place over the full extent of the mix. On the other hand, if the tracer is fed into the mid position (2), equalisation of its concentration only has to take place across one half of the mixer's length, reducing the mixing time from 150 to 40 seconds and thus to about a quarter.

Thus, the way it is loaded considerably affects the mixing time. If the ingredients are poured into the machine in an uncontrolled fashion premixing will already have taken place and the mixing time is significantly shorter than if the components had been put into the machine so that they were spatially separated and one ingredient predominated in certain areas within the mixer.

In mixers like the horizontal mixer of Fig 7.2 which is equipped with fast rotating mixing tools the radial mixing is achieved very fast, in contrast to the mixing perpendicular to the rotation. Axial transport determines the mixing time.

If the tracer is positioned at the borders of the mix this appreciably lengthens the period than if it had been placed in the middle. Consequently the relevant initial conditions when the machine is started form part of the data on mixing times.

Similar to the case in Fig. 7.1 a criterion can also be laid down here for determining the mixing time more precisely. If the high frequency portions are filtered out of the concentration curve the mixing time can be defined as that corresponding to the point at which the concentration first reaches 95 or 105% of the stationary end value μ..

The concentration curves illustrated in Fig. 7.4 only enable limited pronouncements to be made as to the mix's quality since the signal is also affected by the movement of the material caused by the plough shares' rotation. However this can be recognised by a frequency analysis and corrected if necessary.

Since premixing appreciably influences the mixing time, when the mixer is being filled the ingredients should be added at the same time and to the same area within the machine so as to shorten the mixing time.

7.3 Coefficient of Mixing for Specifying Mix Speeds

As already mentioned, detailing a mixing time only has some meaning if the initial conditions are detailed at the same time. To get a dimensionless representation Entrop [23] multiplies the mixing time by the rpm (or to be more exact the rotational frequency) of the mixer's agitators or container. This is a general practice in the field of blending. Entrop produces the following correlation for conical screw mixers whose screws have a rotational frequency of n_s, an immersion depth l_s and a diameter d_s of the screw:

$$t \cdot n_s = 13\left(\frac{l_s}{d_s}\right)^{1,93} \tag{7.1}$$

There is a square relationship by approximation between the dimensionless mixing time and the screw immersion depth l_s (Fig. 7.5). The screw immersion depth determines the mixer's maximum filling level. With this kind of conical screw mixer the concentration's equalisation over the height of the mix determines the mixing time.

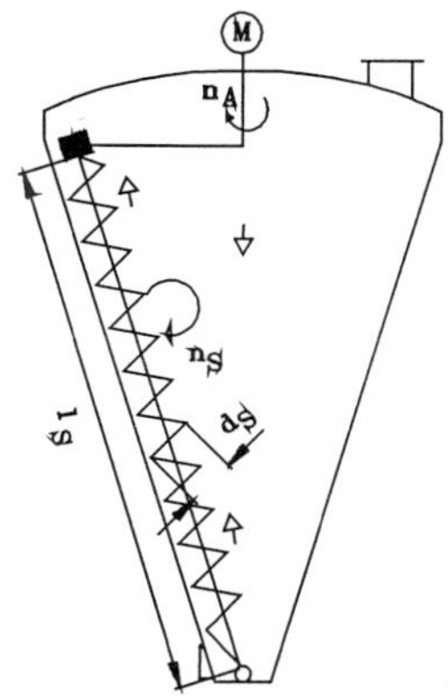

Fig. 7.5 The mixing time in a conical screw mixer is dependent on the screws rotational frequency n_S, the immersion depth l_S and the screw diameter d_S (cf equation 7.1)

However as soon as machines such as fluidised bed or bunker mixers are used it makes little sense to use rotational frequency to set a norm for mixing times. Likewise the initial conditions still have to be detailed at that same time. Consequently Müller [26] suggests using **coefficients of mixing** to specify the mixing speed. The **coefficient of mixing M** is the parameter of a semi empirical diffusion model obtained by comparing experiments and models. The model's basic features are sketched out in the following example of longitudinal mixing in a horizontal mixer.

Fig. 7.6 shows a diagrammatic view of a horizontal mixer with a length L, local coordinate z and rotational frequency n. Müller applies this model in his dissertation [24, 25]. In this type of mixer, gradient equalisation occurs very much more slowly horizontally than radially. Thus the axial concentration equalisation determines the mixing time. In the individual z planes vertical to the mid axis it is assumed by way of a simplification that only an average concentration x exists and from this simplification is produced a one dimensional view.

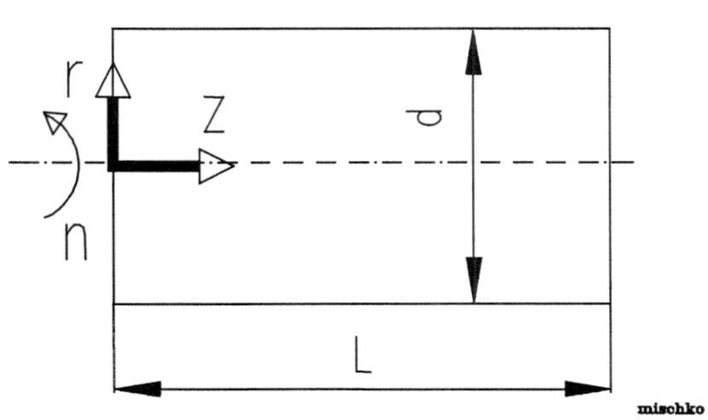

Fig. 7.6 One dimensional model of a horizontal mixer. Axial concentration equalisation in direction z occurs much more slowly than radial (r) equalisation

Unsteady concentration equalisation x(z, t) progresses as described in Fick's second law:

$$\frac{\partial x(z,t)}{\partial t} = M \frac{\partial^2 x(z,t)}{\partial x^2} \tag{7.2}$$

The coefficient of mixing M is constant over the whole mixing area (length z) and at all points of time t. Separation is noten taken into account! Müller gives the following solution for the variance's temporal course under the initial conditions as specified above and where the tracer is next to one of the end walls:

$$\sigma^2(t) = \sigma^2_{syst}(t) + \sigma^2_z \tag{7.3}$$

The following simple relationship is produced for systematic variance so long as the tracer concentration is a great deal less than 1:

$$\sigma^2_{syst} = \mu^2 2e^{\frac{(-2)\pi^2 Mt}{L^2}} \tag{7.4}$$

A large coefficient of mixing M gives a short mixing time. A long mixer length L and thereby a large mixing track has exactly the opposite effect and the mixer length in equation 7.4 is even squared. Equations 7.3 and 7.4 enable the coefficient of mixing M to be determined by experiment. The logarithms of the random sample variances obtained from the trials are taken and plotted across the mixing period (see Fig. 7.7). These logarithms alter equation 7.3 as follows:

$$\ln\left(\sigma_t^2\right) = \ln\left[\left(\sigma_z^2\right) + 2\mu^2 e^{\frac{-2\pi^2}{L^2} \cdot M \cdot t}\right] \tag{7.5}$$

Where there is a short mixing period systematic variance σ^2_{syst} is very much larger than the variance in stationary condition σ^2_z, from which there follows equation 7.7:

$$t \ll t_* \Rightarrow \sigma_z^2 \ll \sigma_{syst}^2 \tag{7.6}$$

$$\Rightarrow \ln\left(\sigma^2(t)\right) = \frac{-4\mu^2\pi^2}{L^2} \cdot M \cdot t \tag{7.7}$$

In the semilogarithmic representation of the variances over the mixing period (Fig. 7.7) the coefficient of mixing M can be determined from the initial slope as shown in equation 7.7.

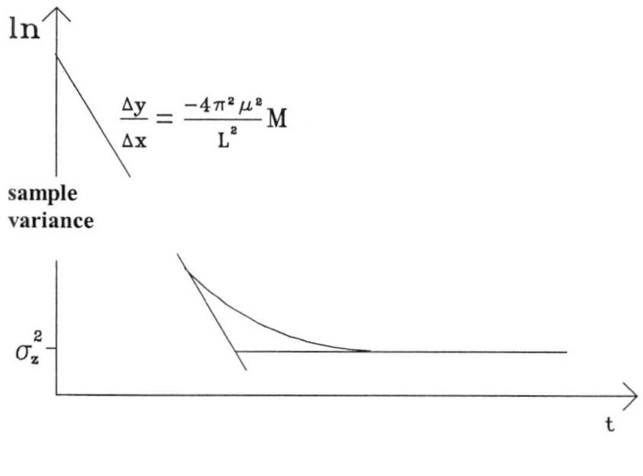

Fig. 7.7 Semilogarithmic representation of the variance over the mixing period. Coefficient of mixing M is established from the straight line slope (separation is not taken into account)

The coefficient of mixing identifies the large scale concentration equalisation which is primarily caused by the movement of large globules towards each other. So it is not surprising that the coefficient of mixing depends in fact on the type of mixer, its geometry and operating conditions and not on the mixture. Müller verified this in respect of solids mixes of differing grain sizes and densities [24, 26]. The modelling as presented assumes that the mix's ingredients do not behave differently in the way they move, thus excluding any tendencies to separation. These can be taken account of in an expanded model by additionally introducing a coefficient of transportation. At this point you are referred to the publications by Sommer [27] and Müller [24]. The dispersion model, which is closely related to the synthetic just illustrated, will be presented in the following chapter on continuous mixing.

7.4 Scaling up Batch Mixers

As with many other processes in which solids play the dominant part, the knowledge base for scaling up batch mixers is meagre. This is because a product's properties can only be recorded with some difficulty. In the latest literary review by L.T. Fan and Yi-Ming Chen [12] the authors recommend three works on the subject of scaling up. The publication by Miyanami [28] is unfortunately in Japanese and another work is the often cited one by W. Müller. In a third book, Rudolf von Rohr and Widmer [29] state that there are still no correlations for ribbon blenders mixers in respect of mixing times. The authors of the literary review therefore come to the conclusion that effective design processes have still to be developed.

Plough share mixers

In scaling up one would like to ensure a *geometric*, *kinematic* and *dynamic similarity* between the pilot and the production mixer [29].

Geometric similarity calls for the same ratios between the linear dimensions in the small and the larger mixer, which is understood, for example, to include the same diameter to length ratios in drum machines. Kinematic and dynamic similarity requires the same speed and power ratios between the two machines. It is easy to see that it is impossible to observe every similarity when scaling up.

Mixing time:

Based on the diffusion model and the results of experiments Müller [26, 24] showed that the following relationship applies to centrifugal mixers:

$$\frac{M \cdot t}{L^2} = \text{const} \tag{7.8}$$

In scaling up it should be well known how mix times change with the design size and rpm of the mixer. The coefficient of mixing M is thereby additionally altered by the Froude number, the mixer's diameter and its filling level. The solids themselves are not an influencing factor in centrifugal mixers since the large scale equalisation which represents the coefficient of mixing is caused by the plough shares.

$$M = f(Fr, d, \text{degree of filling}, \ldots) \tag{7.9}$$

The Froude number includes the rotational frequency and denotes the relationship between centrifugal and gravitational acceleration (cf 6.1). The coefficient of mixing increases as the rotational frequency and therefore Froude number mounts. Müller gives the following relationship for the influence of diameter d on the coefficient of mixing M for standard centrifugal mixers:

$$\begin{aligned} Fr < 3: &\quad \frac{M}{d^2 n} = \text{const} \\ Fr > 3: &\quad \frac{M}{d^2 n} \sim Fr^2 \end{aligned} \tag{7.10}$$

Equation 7.10 produces the following relationship for mixing times:

$$\begin{aligned} Fr < 3: &\quad t \sim \left(\frac{L}{d}\right)^2 \frac{d}{v} \\ Fr > 3: &\quad t \sim \left(\frac{L}{d}\right)^2 \frac{d^3}{v^5} \end{aligned} \tag{7.11}$$

Particularly in the higher Froude number range, faster mixing times are achieved by mixers with a short length and large diameter than by those having the same capacity but a longer mixing length. Scheuber [69] has also carried out investigations into mixing processes in plough share mixers. He used machines with a capacity of 50 to 3000 litres. Meili has examined 150 to 600 litre capacity GMS Multi Flux mixers [72].

Energy consumption:
Energy consumption is determined by the power draw P and the mixing time t_*:

$$E_{mixing} = P \cdot t_* \tag{7.12}$$

The power is required to overcome the resistance caused by the friction of the goods to be mixed. Similar to the practice in stirring, the power draw P is dimensionlessly represented as a Newton number Ne:

$$Ne = \frac{P}{\rho_s(1-\varepsilon)d^5n^3\frac{L}{d}} \tag{7.13}$$

At low velocities (Coulomb friction) frictional forces are not influenced by the speed, hence the following holds for Froude numbers smaller than 3 [26]:

$$Ne \sim \frac{1}{Fr} \tag{7.14}$$

The above relationship applies to centrifugal mixers. The forces of mass gain in significance as the Froude number increases and there are deviations from relationship 7.14. As opposed to the Reynolds number for fluid mixes the Froude number does not describe movement status [26, 30] and the absolute value of the power draw can vary distinctively according to the shape of the agitators. Likewise the solids' roughness and fineness considerably influence the power draw.

Conical screw mixers (see fig. 7.5)
Entrop's correlation [23] regarding power draw includes the screws depth of immersion l_s and diameter d_s as the most important value:

$$\frac{P}{n_s\rho_s(1-\varepsilon)d_s^4g} = k_1\left(\frac{n_s}{n_a}\right)^m\left(\frac{l_s}{d_s}\right)^{1,7} \quad ; m \to 0 \tag{7.15}$$

The correlation for mixing times has already been given in equation 7.1.

Fluidized bed mixers:
Fluidized bed mixers (see Fig. 7.8) represent equipment with complex flow processes both in respect of the solid and gas phase and which processes are still not completely understood. Fluidized bed mixers are designed for mixing very large quantities. The specific energy requirement E_m still cannot be calculated in advance. Equation 7.16 shows some of the essential influencing factors [26]:

$$E_m \sim \frac{z}{M}d_p^2H^2 \tag{7.16}$$

The power requirement is very high in the case of very large particles and likewise increases quadratically with the height of the silo. The relationship between the coefficient of mixing M and the fluidizing status, here represented as the multiple of the release velocity z, still is not clear. Depending on the fluidising velocity coefficients of mixing can amount to a few or several thousand cm/s^2. The mixing speed rises at higher silo diameters. With this type of mixer one dimensional models are pushed to their limits since axial and radial mixing processes have to be taken into account in any accurate assessment.

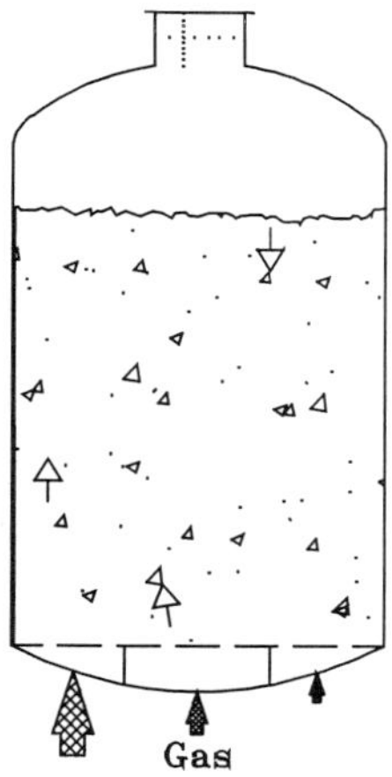

Fig. 7.8 Principle of a fluidized bed mixer

Since movement is highly dependent on the solid's fluidizing properties it is no wonder that this also applies to coefficients of mixing. Exchange processes and hence also the coefficient of mixing alter across the height of the fluidized bed (see [31] - [33]).

8 Continuous Mixing of Solids

In a continuous mixing process the ingredients are continuously fed into the mixer then mixed and the product prepared for the next processing stage. The operations of feeding, mixing and discharging follow each other locally but contemporaneously.

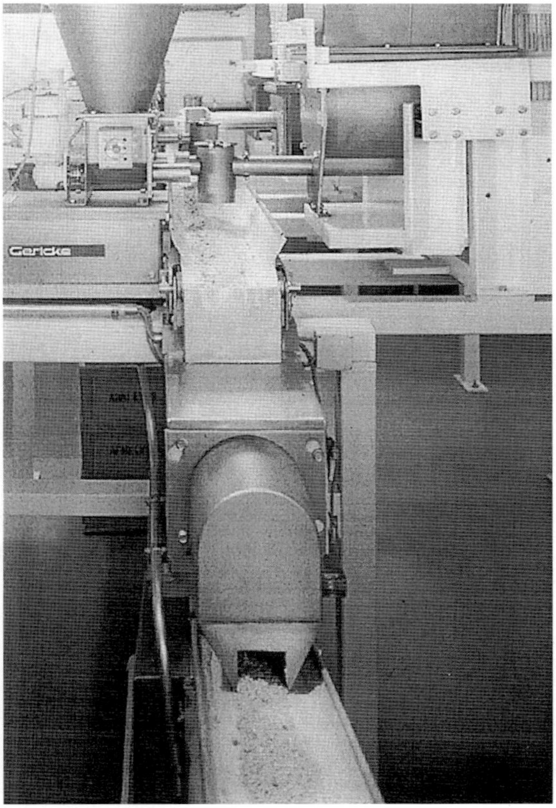

Fig. 8.1 Production of Muesli: Solids feeder supply at constant rate the components (raisins, flakes, etc.) onto a belt which deliver the components to the continuous mixer (bottom of the picture). The continuous mixer discharges onto a second belt.

As opposed to batch mixing, the feeding process in a continuous system appears to be more costly. In continuous mixing the weighing and filling of a batch mixer is replaced by the ingredients' controlled continuous addition. This is particularly expensive when producing multi-ingredient mixtures. The blending time in a continuous mixer is laid down for a given throughput by the material's residence time in that machine. Losses of product during start up or shut down added to this lower degree of flexibility come as further disadvantages of the continuous process. Yet it possesses considerable advances over batch processing both in financial terms and in respect of process control:
Even high throughput continuous mixers are compact. This means short mixing paths and thus altogether easier mixing. When integrated into a continuous production system a continuous mixing process saves on reservoirs or silos and automating the course of the process is simplified. In the case of dangerous products or base materials there is less potential risk with a continuous process since only a small quantity of material accumulates in the mixer. Segregation can be limited in a continuous mixer. It was pointed out in Chapter 2 that free flowing solids tend to segregate as soon as there is a difference in the way the mixture's individual components move. Segregation occurs when a silo is being emptied, the mix is transported or as soon as a batch mixer is being discharged. A continuous mixer, which on account of its compact construction can be positioned before the next station in the processing chain, guarantees that a mix of a higher quality will in fact be made available to that next stage of the process.

The continuous mixer has principally two tasks (Fig. 8.2): The ingredients, which in an extreme case arrive in the mixer side by side, have to be **radially** mixed (r). In this case radial means lateral to the direction of the material's conveyance into the mixer. If feeding is not accurate and at constant rate or the ingredients are themselves unhomogenized the mixer has also to equalise any differences in concentration in an **axial** direction (z), i.e. in the direction of the material's conveyance.

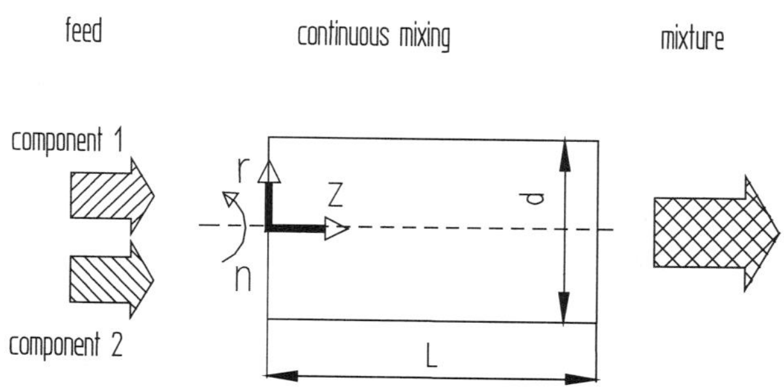

Fig. 8.2 The continuous mixing of two ingredients

If a mixer only has to perform its task radial it can have a very compact structure since slim line mixers with a high rpm very quickly equalise concentrations radially over short mixing paths. The residence time of the material in the mixer has to be just long enough to reliably deliver this. Here one has to check to what extent radial mixing is at all necessary or whether time constant feeding of all the ingredients, whereby the feeder streams are brought closely together (Mixing in the Feed Stream, Chap. 6), will suffice for the application in question.

8.1 Variance Reduction as a Yardstick for Axial Mixing in Continuous Mixers

It was shown in Chapter 3 that the quality of a mix can be described using its variance or power density spectrum. Naturally this also applies to products leaving a continuous mixer. Besides the quality of a continuously produced mix, it is also of great interest to what extent by axial mixing a continuous mixer can remove inconsistencies in the ingredients being fed in or fluctuations in the feeding. Dankwerts [44] put forward the term **"variance reduction ratio"** or VRR to describe this extent. It is defined as the ratio of the variances at the mixer's inlet and outlet:

$$\mathrm{VRR} = \frac{\sigma_{\mathrm{in}}^2}{\sigma_{\mathrm{out}}^2} \tag{8.1}$$

Thus the variance reduction ratio compares the mix quality at the machine's inlet and outlet and thereby distinguishes the efficiency of a continuous mixer. This can be illustrated using a practical example:

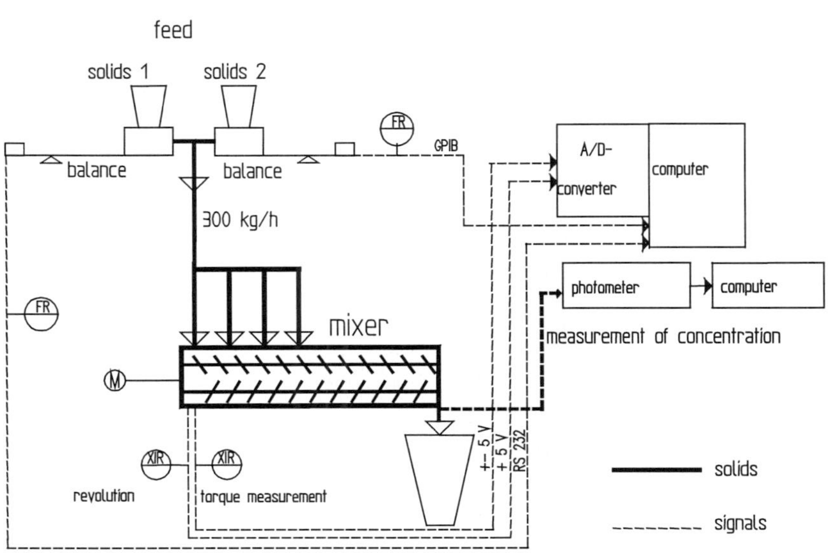

Fig. 8.3 Diagram showing the continuous solids mixing test plant at the Federal Institute of Technology (ETH) Zurich [30]. The tracer element's input concentration is established through signals from the scales and the concentration curve at the mixer's outlet is obtained by taking fibre optic in-line measurements.

Fig. 8.3 shows the diagram and 8.4 a photo of the test installation at the Federal Institute of Technology (*ETH*) [30] for examining continuous mixing. Two feeders conduct the ingredients into the mixer. Both feeders together with their bins stand on scales and the flows and concentration of the tracer element at the mixer's inlet are determined by subtracting the weight over the test period. Once again the principal ingredient is white $Al(OH)_3$ which is mixed with the dark tracer SiC. This occurs in a horizontal twin shaft Multiflux II (Gericke) mixer. The fibre optic probe described in Chapter 5 and which enables the SiC concentration to be ascertained in-line is installed at the mixer's outlet.

Fig. 8.4 Photo of the continuous solids mixing test plant at the Federal Institute of Technology (ETH) Zurich [30].

In order to establish the variance reduction ratio in this machine, feeding was deliberately impaired by periodically switching the SiC tracer element on and off. Since the selected tracer concentration is very low the level of product in the mixer is just about constant. The SiC tracer concentration at the mixer's inlet is shown in Fig. 8.5a as a line of dashes. The measurement values were calculated from the signals from the scales. The tracer concentration at the mixer's outlet, which were ascertained using the fibre optic probe, are shown in Fig. 8.5a as a solid line.

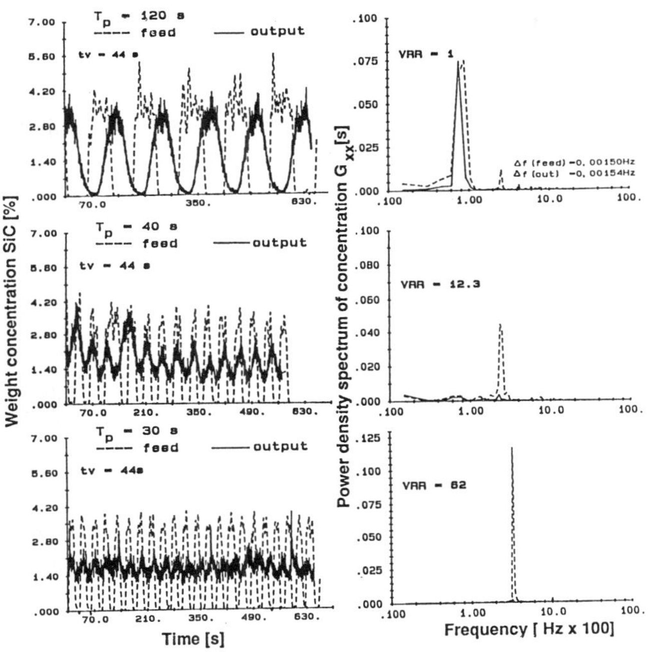

a) b)

Fig. 8.5 Variance reduction ratio **VRR** in continuous Multiflux II solids mixer, SiC-feed oscillating with different periods T_p, $Al(OH)_3$ feed at constant rate (20g/s) mean residence time t_V=44 s

a) variation in time of SiC concentration

b) power density spectrum of SiC concentration

Variance reduction ratio: rough analysis of the test series in Fig. 8.5a

For the purposes of an initial analysis of the series of tests we shall focus on the three concentration sequences in Fig. 8.5a at the mixer's inlet and outlet. By periodically switching the feeder on and off the concentration at the inlet fluctuates in the form of a rectangular oscillation. Since fluctuations (deviation) around a mean value signify a high variance, the variance of the input is high.

The mixer should now dampen these initial oscillations, thereby reducing the variance. In the top test sequence in which the tracer feeder is switched on and off every 60 seconds, thus following a 120 second cycle, this does not happen. The fluctuation at the mixer's outlet is as great as at its inlet. This is not astonishing, since the average residence time in the mixer is only 44 seconds and therefore plainly less than the feeder oscillation cycle. The cycle was shortened in the next two tests. Interval T_p was then 40 and 30 seconds respectively, with the average residence time in the mixer being once again 44 seconds. The mixer was now able to clearly reduce these input fluctuations.

The variance reduction ratio, the mixer's capacity to dampen input fluctuations, is therefore essentially determined by the ratio of the feed fluctuation cycle to the average residence time in the mixer.

Variance reduction ratio: detailed analysis of the test series in Fig. 8.5a

In the series of tests shown in Fig. 8.5, with fluctuations around an average concentration and feeding interval, it is natural to apply the methods of frequency analysis to the test results. The power density spectrum which has already been presented to the reader in Chapter 3 was applied to the mainly periodic measurement data in Fig. 8.5. The power density spectrum represents an elegant method for determining the variance reduction ratio.

The feeding fluctuations in Fig. 8.5a are rectangular oscillations with cycles T_p of 120, 40 and 30 seconds. A rectangular oscillation is mathematically a superimposition of several sinusoidal oscillations (Fourier synthesis) [49]. The basic oscillation has a cycle T_p. To this are added oscillations with shorter cycles ($T_p/3$; $T_p/5$; $T_p/7$; ....) with a markedly smaller amplitude. In Fig. 8.5a above, the mixer dampens the higher frequency oscillations (= those with a shorter cycle) and this shows in the loss of their rectangular profile. The concentration at the mixer's outlet fluctuates in an approximately sinusoidal fashion. However, the basic oscillation having a cycle T_p=120 seconds appears unchanged and has the same amplitude at the mixer's outlet. Amplitudes, like variances, describe the scatter around an average value and both the amplitude of this basic oscillation and the variance were not reduced in the mixer (VRR=1).

In Fig 8.5b the frequency analysis of the concentration curve is shown by means of the power density spectrum. Sinusoidal fluctuations in period T_p possess a peak at point $1/T_p$ in the power density spectrum. If this peak is multiplied by the sampling frequency (corresponding to the area) you get the variance at the relevant frequency (cf. Chap. 3). The variance reduction ratio is then obtained as shown in equation 8.2.

$$\mathrm{VRR}(f) = \frac{(\mathrm{peak}(f)\cdot\Delta f)_{\mathrm{input}}}{(\mathrm{peak}(f)\cdot\Delta f)_{\mathrm{output}}} \tag{8.2}$$

Our further observations will concentrate on the basic oscillation with a period T_p. If the period for feeding the trace ingredient is shortened (Fig. 8.5, centre and bottom) to 40 and 30 seconds respectively, dampening of the basic oscillation occurs in the form of peak reduction. There is already considerable dampening at a period of 40 seconds and the variance reduction ratio reaches the value of 12.3. At a period of 30 seconds there is only the suspicion of a peak at the mixer's outlet and the variance reduction ratio produces a value of 62. These trials show very graphically the importance of time constant feeding in continuous processes.

Time constancy in the feeding of solids has increased in the last few years.

There follows an overview of continuous feeding and mixing in Chapters 10-11. High long term constancy is achieved by using gravimetric feeding systems, e.g. loss-in-weight-feeders. Specially designing the feeder unit reduces the occurrence of cohesive powders falling from it in clumps. If the ingredients are only added in very small quantities continuous, controlled feeding is made more difficult on account of the limited resolution of the flow rate measurements. Cyclic feeding offers itself whereby the mixer then also has to equalise concentrations axially as well. An extremely short time constant feeding principle for throughputs of several tons an hour has been developed by Tesch and Reh [45]. Free flowing solids are fed in through a nozzle at the bottom of a stationary fluidised bed. Feeding was constant even at fractions of a second.

The variance reduction ratio achievable in a mixer is determined along with the feeding's quality by the *residence time* and *residence time distribution* of the material in the machine. These terms are used in every continuous process and are consequently explained in the following.

8.2 Residence Time Distribution: Definition and Determination by Experiment

An axial mix can only come about if the time within which the particles are transported through the mixer is no longer uniform but fluctuates around an average value - the mean residence time.

In the case of continuous chemical reactors - and mixers frequently perform additional material modification tasks (reaction, drying, granulation etc.) besides mixing - an axial mix is for the most part undesired since there is a drop in yield. However a low yield at the same throughput means a larger mixer (reactor). With axial mixes the danger of segregation is further increased. A narrow residence time distribution of the solids in the mixer guarantees that stresses upon the product are uniform. Therefore following the ideals of process control a continuous mixer should not be presented with the additional task of equalising fluctuating concentrations of the ingredients as a result of non uniform feeding.

In a continuous flow system such as a mixer (or reactor, mill, grader etc.) the particles which entered the mixer at the same time do not leave it at the same moment since each takes its own individual, stochastic path through the machine. A residence time distribution occurs. Bourne [37] defines residence time as follows: *The residence time of an ingredient in the mixer is the time during which the ingredient remains inside it.*[1]"

Residence time is a random value since the movement of the individual particles in the mixer is subject to chance. It is just this random nature of the particles' movement which leads to no constant feeding conditions in the mixer

1 When talking about residence time in a solids mixer it is presumed that the ingredients are not being transported in the mixer individually but as a mixture. This assumption needs to be checked by means of experiments in the case of segregating products in particular.

being dampened. Thus a broad residence time distribution signifies extensive axial mixing.

Residence time is defined mathematically as follows: if **E(t) dt** denotes the proportion of the flow exiting from the mixer after a residence time of t and t+dt and **F(t')** denotes that of the flow with a residence time of <t′ then:

$$F(t') = \int_0^{t'} E(t)dt \tag{8.4}$$

t is a distributed variable, E(t) is the density function and F(t) the residence time's cumulative distribution function (Fig. 8.6). Thus it follows from equation 8.3 that:

$$E(t) = \frac{dF(t)}{dt} \tag{8.4}$$

As with other distribution functions, e.g. particle size, you get characteristic numbers by calculating their moments:

$$\int_0^{\infty} E(t)dt = F(\infty) - F(0) = 1 \tag{8.5}$$

In stationery operating conditions the **mean residence time t_v** corresponds to the ratio of the accumulated volume in the mixer to the throughput:

$$t_V[s] = \frac{\text{product accumulated in the mixer}[kg]}{\text{mass flow rate}[kg/s]};\ \rho_s = \text{const} \tag{8.6}$$

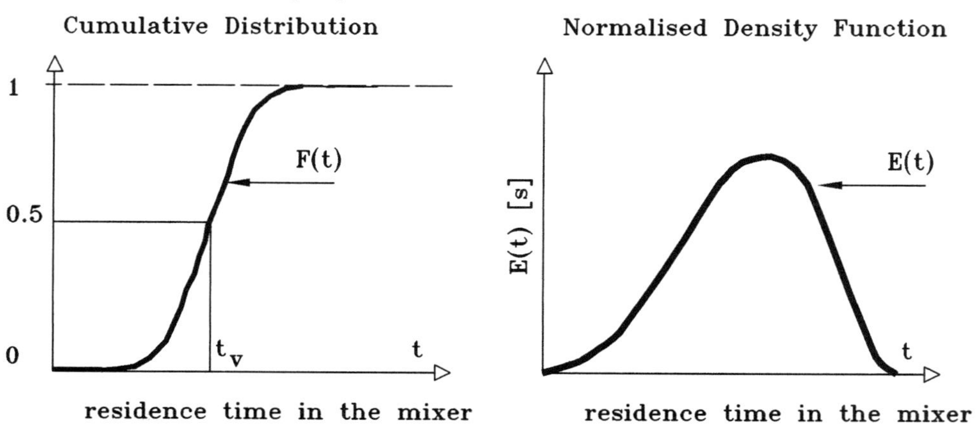

Fig. 8.6 Residence time distribution represented as the cumulative distribution (F(t) and density distribution (E(t)

The average residence time t_v can be derived from the residence time distribution by calculating the 1st moment:

$$t_V = \int_0^\infty tE\,dt = \int_0^1 t\,dF \tag{8.7}$$

The variance is a yardstick for measuring the breadth of the residence time distribution and thereby the axial mixing. It is calculated using the 2nd primary moment:

$$\sigma_{t_V}^2 = \int_0^1 (t-t_V)^2 dF = \int_0^\infty (t-t_V)^2 E\,dt \tag{8.8}$$

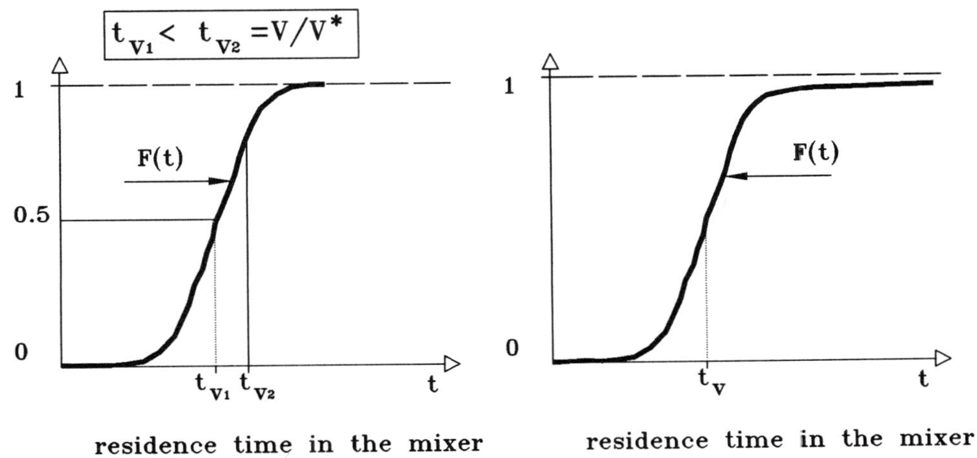

Fig. 8.7 The residence time distribution shows anomalies when the material is being transported through a continuous mixer

The residence time distribution represents the temporal course of the material's passage through the mixer and is revealed by poor flow guidance in the machine. Two examples of this are to be found in Fig. 8.7. In the left hand illustration an average residence time t_{v2} had been expected but in fact the measured average residence time t_{v1} is plainly shorter. Half of the material is transported very quickly through the mixer and the other half does not take part in the mixing process and remains stationery in the machine. This leads to a reduction in the average residence time. In the illustration on the right one can see that part of the material takes a very long time to leave the mixer. This "tailing" points to "dead zones" in the machine which can greatly delay the material and make it difficult for it to exit from them. A detailed description of these effects is given by Levenspiel [42] and Villermaux [73].

The residence time distribution is determined by imposing an interference factor on the mixer whilst it is in stationery operation and observing how this interference is broken down inside it (the stimulus/response method). The addition (or removal) of a tracer produces the interference and the tracer's concentration is measured at the mixer's outlet. In the case of solids it is easier to perform a jump test than e.g. using the pulsed addition of a tracer. An overview of other interference signals is given by Wen and Fan [38]. A traditional jump test is shown in Fig. 5.8 in Chapter

5. The cumulative residence function F(t) similar to that shown in Fig 8.6 is obtained from this jump test by averaging the concentration curve x(t) with the initial concentration x_1 and the final concentration x_2.

$$F(t) = (x(t) - x_1) / (x_2 - x_1) \tag{8.9}$$

The tracer must be quantitatively measurable in small concentrations in order not to change the flow pattern. Its motion should correspond to that of the mixture. Holzmüller [10] and Fan [40] use radioactive tracers to determine the residence time distribution. The expense and safety requirements for this technology are nowadays enormous. Weinekötter [30] uses dark SiC or green Irgalite as a tracer, whose concentration in a mixture of white $Al(OH)_3$ can be optically measured using the procedure described in Chapter 5.

8.3 Description of Axial Mixing using the Dispersion Model

The velocity field in a mixer is very complex and it is still not possible to describe it theoretically. Nevertheless in order to depict a mixer's flow behaviour one relies on empirical models and tries with the help of these models' parameters to record the mixer's actual behaviour.
The **dispersion model** is a classical one. It is derived from the general transportation equation and is based on the following concept (Fig. 8.7).

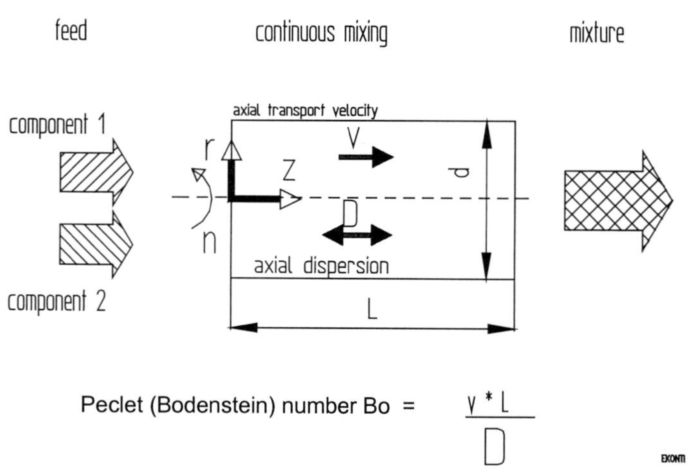

Fig. 8.8 Schematic illustration of the dispersion model

The movement of solids in the mixer can be described by overlaying a plug flow with velocity v with a random dispersion process with a coefficient of dispersion D. Dispersion is imposed on the solid by the elements in the mix. Its effect is that the solid does not pass through the mixer at a uniform speed (like the flow through a pipe). The greater the coefficient of dispersion D the greater the axial mixing. Several suppositions are made in formulating the *general transportation equation* [41] (Fig. 8.8):

- The mixer's longitudinal axis is the z axis
- The material's entry and exit are through the end faces
- The cross sectional area is constant across the length
- There is no heterogeneity in a radial direction (one dimensional model; L>>d)

The general transportation equation is then written as follows [38]:

$$\frac{\partial x}{\partial t} = -v \cdot \frac{\partial x}{\partial z} + D \cdot \frac{\partial^2 x}{\partial z^2} \tag{8.10}$$

D denotes the axial coefficient of dispersion (compare this with the coefficient of mixing in batch mixing, Chapter 7). Equation 8.10 is usually represented in a dimensionless form. To do this the relevant sizes are averaged by the mixer length L and the average residence time t_v.

$$\xi = z/L;\ \theta = t/t_v;\ t_v = L/v \tag{8.11}$$

The **dimensionless form of the general transport equation** then follows from equations 8.8 and 8.9:

$$\frac{\partial x}{\partial \theta} = \left[\frac{D}{v \cdot L}\right] \cdot \frac{\partial^2 x}{\partial \xi^2} - \frac{\partial x}{\partial \xi} \tag{8.12}$$

The expression in brackets deserves particular attention:

$$\frac{v \cdot L}{D} \equiv \text{Bodenstein number} \equiv \text{Peclet number} \tag{8.13}$$

The **Peclet number** (Bo) denotes the ratio of axial convective velocity v to the coefficient of dispersion D (Fig. 8.8). If the Peclet number is close to zero dispersion preponderates. The residence distribution time in the mixer approaches that in an ideal **agitated (stirred) vessel** [42]. If the Peclet-number is very large there is a straight **plug flow**.

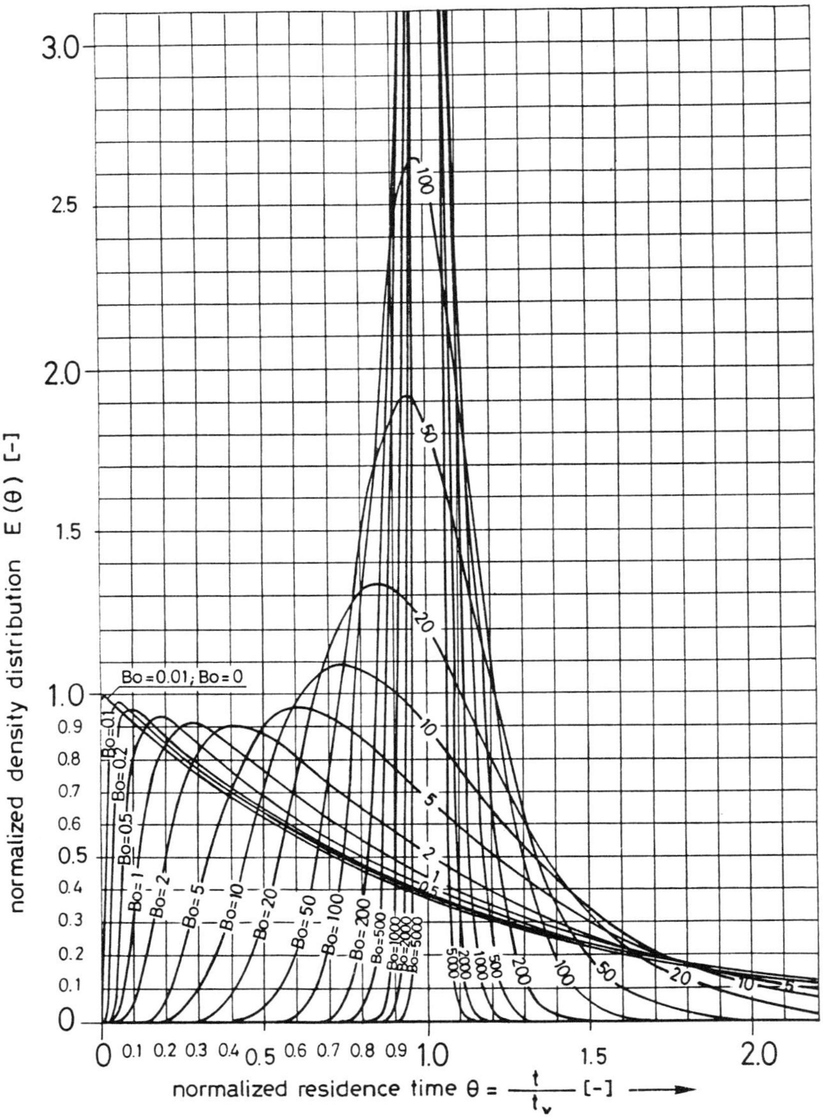

Fig. 8.9 Residence time distribution as a function of the Peclet number (Bo) [43, 30]

In order to apply equation 8.12 the boundary conditions have to be clarified. It is presumed that axial dispersion is only present inside the mixer and that D = 0 before the material enters and after it leaves the machine. Levenspiel terms this a "closed vessel" [42]. However, using these boundary conditions equation 8.12 cannot be solved analytically and a numerical solution is required. The result of the numerical solution of the general transportation equation provides the residence time distribution for various Peclet numbers [30] (Fig. 8.9). The density distribution is plotted as a function of the averaged residence time (E = E(Θ); eq. 8.11). Where Θ equals 1 the dimensionless residence time corresponds to the average residence time. For small Peclet numbers the residence time distribution becomes very wide and asymmetric. With an increasing Peclet number the residence time distribution becomes narrower and finally approaches that described by a plug flow behaviour of a pipe reactor.

Axial mixing in a reactor with a flow passing through it is frequently modelled as a cascade of agitated vessel. However, in the case of a high degree of axial mixing there are significant differences between the stirred vessel cascade and dispersion models. The two models adapt as a flow takes on an increasingly purely transportation character. The following relation can be made between the parameters of both the above models, the Peclet number = Bo (Bodenstein) and the number of agitated vessels (N_R)

$$Bo = 2 \cdot N_R \quad ; \quad Bo, N_R \to \infty \tag{8.14}$$

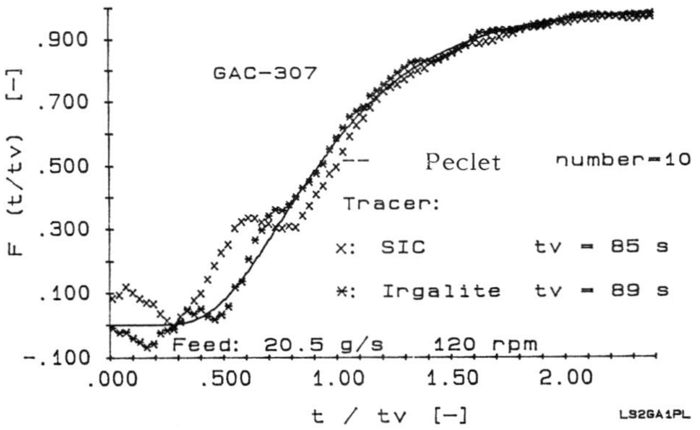

Fig. 8.10 Cumulative distribution F(t) of the residence time in the GAC-307 mixer: Dispersion model and experiments
Mixture: $Al(OH)_3$ ($d_{p3.50}$=70 μm) with SiC ($d_{p3.50}$=26 μm) or Irgalite ($d_{p3.50}$=0.22 μm)
Average residence time t_v = 85 and 89 seconds

If the residence time distribution in a mixer has been measured e.g. by a jump test for certain operating conditions this residence time distribution ascertained by experiment can be compared with a residence time distribution calculated using a dispersion model. Here the Peclet number will vary until the residence time distribution from the dispersion model corresponds to that obtained experimentally. This is illustrated in Fig. 8.10 which shows the cumulative residence time distribution in a small continuous mixer (type GAC307, Gericke). From a comparison of the experimental points with those using the dispersion model it follows that under these operating conditions this machine can be described very well by a Peclet number of 10 [30].

8.4 Diagram for Designing a Continuous Mixing Process for Solids

The experiment shown in Fig. 8.5 illustrates very vividly that the variance reduction ratio is determined by the constancy with which the solids are fed and by their residence time distribution in the mixer. Dankwerts [44] showed that the Variance reduction ratio VRR can be calculated if you know the **autocorrelation function $R_{xx,feed}$** of the property in the feeder flow to be examined, which should be as homogeneous as possible, and the **residence time distribution E(t)** of the product in the mixer:

$$\frac{1}{VRR_{fluid}} = 2\int_{\tau=0}^{\infty}\int_{t=0}^{\infty} R_{xx,feed}(\tau)E(t)E(t+\tau)dtd\tau \qquad (8.15)$$

The autocorrelation function $R_{xx.feed}$ marks the quality of the feeding and is calculated in accordance with the following equation:

$$R_{xx.feed} = \lim_{T\to\infty}\frac{1}{T}\int_{t=0}^{T} x_{feed}(t)\cdot x(t+\tau)dt \qquad (8.16)$$

Equation 8.15 does not take separation processes into account. Wang [48] applies this formula amongst others to the continuous mixing of solids. Williams [3] also considers equation 8.15 suitable in principle for forecasting the variance reduction ratio in the continuous mixing of solids. In addition he proposes a related method which likewise uses residence time distribution [46]. Fluctuations in the concentration of the tracer in the feed flow are multiplied by each other in a matrix. Equation 8.15 serves as the basis of a configuration diagram for continuous mixing processes [30] and for this we go to a dimensionless description. To that end the actual residence times in 8.15 are averaged by the mean residence time t_v:

$$\frac{1}{VRR_{fluid}} = \frac{\sigma^2_{input}}{\sigma^2_{output}} = 2\int_{\tilde{\tau}=0}^{\infty}\int_{\Theta=0}^{\infty} R_{xx,feed}(\tilde{\tau})E(\Theta)E(\Theta+\tilde{\tau})d\Theta d\tilde{\tau} \qquad (8.17)$$

In the previous chapter axial mixing in a continuous mixer was described using the dispersion model where the residence time distribution E arose as a function of the Peclet number (Fig. 8.8).

$$E(\theta) = f(Bo) \qquad (8.18)$$

Strictly speaking, Dankwerts' equation 8.17 for calculating the variance reduction ratio only applies to the continuous mixing of fluids. In the case of solids it must be taken into account that the minimum value acceptable by the variance, which indeed describes the mixture's homogeneity, is limited. Consequently Dankwerts' formulation has to be supplemented by an additive factor which takes the final homogeneity of solids mixtures into consideration:

$$\frac{1}{VRR_{solids}(\hat{f})} = \frac{\sigma^2_{end,solids}}{\sigma^2_{input}(\hat{f})} + 2\int_{\tilde{\tau}=0}^{\infty}\int_{\Theta=0}^{\infty} R_{xx,feed}(\tilde{\tau})E(\Theta)E(\Theta+\tilde{\tau})d\Theta d\tilde{\tau} \quad ; \hat{f} = \frac{t_v}{T_p} \tag{8.19}$$

$\sigma^2_{end,solids}$ represents the variance of a random mix in an ideal case and is thereby dependent on the concentration, sample size and particle size distribution. For real mixes $\sigma^2_{end,solids}$ represents the variance appearing at the mixer's outlet under a stationery condition at an ideal feeding at constant rate.

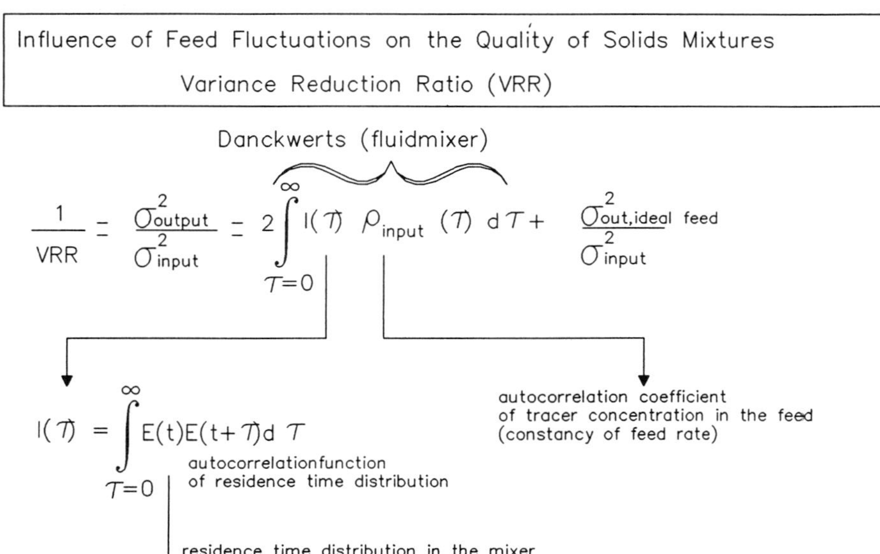

Fig. 8.11

In addition the type of mixing task (and therefore $R_{xx,feed}$) must be specified. The mixer has to equalise a sinusoidal fluctuation in the tracer concentration where the ratio of the feed fluctuation period T_p to the average residence time t_v in the mixer varies over a wide range. In the case of a sinusoidal fluctuation of the tracer component in the feed the input variance is unrelated to the interval duration and is shown as:

$$\sigma^2_{input,sinus}(\hat{f}) = \sigma^2_{input,sinus} = \frac{1}{2}\cdot\mu^2 \tag{8.20}$$

We have thus presented all the modelling concepts dealing with the configuration diagram. Fig. 8.11 illustrates the various models combined together.

Fig. 8.12 shows the result of the model for two mixes. The feed fluctuation of the tracer concentration, which has been averaged by the mean residence time t_v, is plotted over the period. The variance reduction is the y axis. The Peclet number is the parameter and thus the degree of axial mixing (back mixing) occurring in the continuous mixer. Also plotted are experimental values obtained from tests similar to those shown in Fig. 8.5.

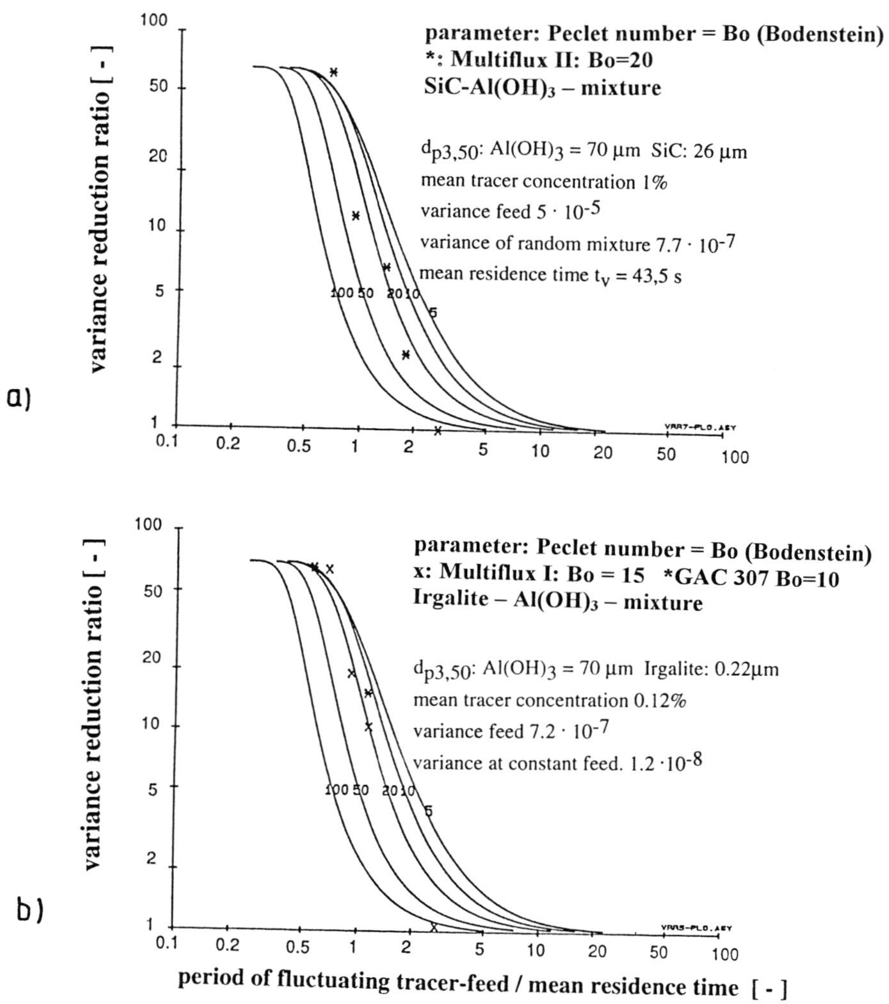

Fig. 8.12 Variance reduction ratio **VRR**: model and experiments
a) SiC/Al(OH)$_3$ mix: Multiflux II mixer **b)** Irgalite/Al(OH)$_3$ mix: Multiflux I and GAC 307 mixers

As the feed fluctuation period lengthens the mixer's ability to reduce the input variance falls off sharply. However, if the feed fluctuation period is noticeably shorter than the average residence time the mixer's ability is enormously increased. If you are working in this range the amplitude of the feed fluctuation is almost inconsequential since there is a high dampening effect.

Summary:
A high variance reduction ratio can be achieved in two ways for a Peclet number:
- Low frequency feed fluctuations with a long period T_p are broken down in high capacity machines with long residence times.
- A feeding system optimised for the mixing process only allows high frequency feed fluctuations enter the mixer. This enables more compact machines to be used.

The final way is preferable from the standpoint of process control technology. The configuration of continuous mixing processes requires consideration of the whole operation. The feed essentially determines the size of a machine and the process's efficiency.

9 Related Processes to Mixing

In many solids mixers powdered ingredients undergo further treatment in addition to pure mixing. For example: agglomeration, disagglomeration, coating, wetting and moistening of the particles as well as heating (or cooling), drying and reactions.

Fig. 9.1 Pressurized atomization: Addition of low viscous liquids to solids by using single phase nozzles (Courtesy of Spraying Systems Inc.)

9.1 The Addition of Fluid Ingredients

Fluid ingredients can be added into the solids mix for various reasons:
- **as an added active ingredient**
- **to bind dust together**

- **to avoid segregation by introducing liquid bridges**
- **to produce or enhance granulation**
- **to coat particles with a layer of liquid**
- **to augment the weight using a cheap ingredient (such as water)**

Fractions of a percent up to 10 percent by weight and more are added in the form of liquids. In general there are two tasks which have to be accomplished:
- The addition and uniform mixing of the fluid with the solids inside the available mixing time in the case of a batch mixer and within the corresponding capacity and residence time in the case of continuous mixing.
- The avoidance of inconsistencies caused by liquid enriched lumps or agglomerates. Mechanical disagglomeration is frequently required, either using the mixer's tools themselves or attached additional high speed delumpers (Fig. 9.3) or both for dispersion of the lumps. There are several ways for adding the liquid component:
- pouring the liquid into the operating mixer filled already with the solids, e.g. lecithin into sugar mixtures
- **atomising the liquid** and direct the droplets onto the particulate components, e.g. adding perfume components to detergents
- pouring the liquid and **dispersing** it within the particles by **high shear or impact forces**, e.g. fat and oil into flour by special fast rotating mixers or mixers equipped with fast rotating delumpers
- Adding the liquid as vapor which condensates while in contact with the particles

Combinations of the above mentioned ways, e.g. atomising and the use of delumpers are frequently used. Trials in smaller pilot plant mixers are helpful for determination of the optimal way of liquid – solids mixing.

Atomisation:
By atomisation fine droplets are produced. If this spray is guided onto the particulate components being moved in a mixer, high quality liquid-solids mixtures can be produced. There are several types of nozzles [74] to perform the spraying: in order to produce fine droplets there has to be a high differential between the relative speeds of the liquid and the gases in the atmosphere into which the liquid is introduced. In *pressurised atomisation* (Fig. 9.1) the liquid is squeezed through an orifice under high pressure. This converts the pressure force into kinetic energy. The sizes of droplets produced is determined by the relative velocities of the fluid and the gas. The fluid flow is proportional to the square root of the pressure and thus only variable within distinct limits (10:1). In *rotary atomisation* the fluid is placed in the centre of a disc or basin, moves to the edge of it through centrifugal force and is finally flung off at a high tangential velocity. Since the speed of rotation and the liquid's flow rate can be adjusted independently of each other, rotary atomisers are highly flexible. Finer droplets than with pressurised or rotary atomisers are produced in 2-phase-nozzles (Fig. 9.2). Here a fast travelling gas is mixed with a relatively slow moving liquid. Such nozzles are divided into

internally mixing and externally mixing. In internal models the liquid and the gaseous atomising medium (mostly air, nitrogen or steam) are fed into the nozzle together. With external mixers (Fig. 9.2) the stream of gas meets the liquid outside of the nozzle at a high velocity (>100 m/s).

Fig. 9.2 2-phase-nozzle (external mixing) for atomising high viscous liquids. The liquid is exiting the nozzle via the central bore. Via the 2 peripheral bores high velocity gas jets hit the liquid jet and disperse the liquid into droplets (Courtesy of Spraying Systems, Inc.)

Here the pressure of the liquid can be distinctly less than that of the gas since in contrast to what happens with internally mixing jets they do not mutually effect each other. Isenschmid et alii [76] are investigating 2-phase-nozzles used for fluids and suspensions. The particle size distribution of the droplets is determined by the liquid's viscosity, the proportion of solids in the suspension and the ratio of the volumes of gas and liquid. An externally mixing jet requires a greater flow of gas than an internal mixer to produce comparably fine droplets [77]. If 2-phase-nozzles are used the gaseous atomising agent will flow into the mixer and has to be removed from the mixing area and released from the particles in special separators.

The aim of atomising is often the production of very fine droplets. However, each atomiser forms droplets whose size falls within a defined range. In most cases what is desired is a narrow droplet size distribution which is also to be sprayed uniformly onto the particles in relation to the surface. As opposed to a lot of other equipment in chemical engineering a mixer can however equalise nonuniform coverage of an area.

Dispersion of liquids into solids by impact or shear

Liquid enriched lumps or agglomerates can be regarded as inhomogeneities. In this case agglomeration is not desirable. Such lumps can be destroyed by impact and

shear forces. Fast rotating mixers (e.g. tpye Schugi) or standard mixers with integrated fast rotating knives are used (Fig. 9.3).
Such dispersion needs high energy input. However, such dispersion the already formed lumps cannot always be avoided. In general, it should be tried to directly produce uniform solids-liquid-mixtures, e.g. by atomising the liquid. The same devices are used for crushing agglomerates or fragile particles.

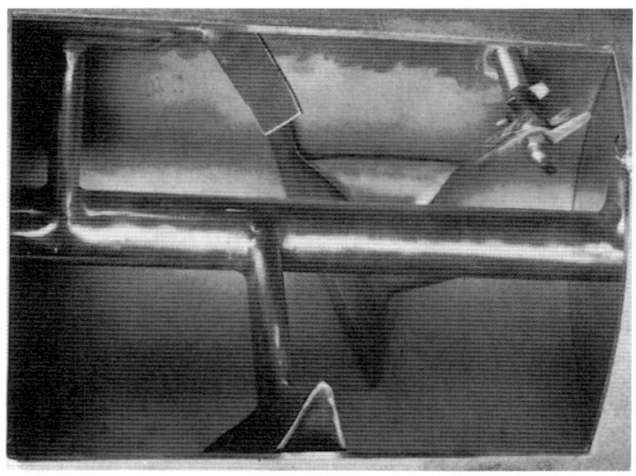

Fig. 9.3 **Horizontal Test Mixer (volume 150 l) equipped with a Delumper**: The mixing tools itself operate with a tip speed of less than 10 m/s. The delumper which consists of a set of fast rotating knives (3000 rpm) operates with a tip speed of above 20 m/s. Larger industrial scale mixers e.g. with a volume 2000 l mixer, are equipped with up to 4 delumpers.

9.2 Growth Agglomeration (Particle Size Enlargement) in Mixers

Many fine disperse solids such as ores and concentrates, the metal industry's iron and non-ferrous deads, cement raw meal, phosphates, fly ash, fertiliser, animal feed, detergents, pharmaceutical products and powdered products of the foodstuffs (instant products) industry are agglomerated.
By agglomeration the following characteristics of solids are positiv influenced:
- increased flowability
- higher bulk density (i.e. lower packaging volume)
- improved wettability e.g. for instant products
- reduced resistance to fluid flow (reduced pressure drop)

Agglomeration also serves to fix the condition of the produced mixture [78].

Sommer [79] sets out 5 agglomeration processes:
1. Growth agglomeration (pelletising or tumble agglomeration)
Fine particles are brought into contact with each other by agitation. In most cases water or other binding agents are added for enhancing agglomeration. The grains are enlarged through capillary forces and more seldom by van der Waals forces (Fig 9.4). Inclined drums, pelletising discs, paddle and plough share mixers and

fluidised beds are typical pieces of equipment.

2. Spray agglomeration

The droplets produced by atomising suspensions are brought into contact with hot gas. This reduces the moisture content in the droplets and pre-dries the agglomerates. These are mostly held together by capillary forces but to this are added bridges formed out of solids at the points of contact. Spray agglomeration is primarily used in the chemicals, pharmaceuticals and foodstuffs (e.g. powdered milk) industries.

3. Selective agglomeration

A further, non mixable fluid phase which moistens the solids is added to a suspension. Interparticle forces cause the agglomerate's formation and coherence. Selective agglomeration can also be used with solids mixtures.

4. Pressure agglomeration

With the addition of a small amount of liquid, solids are agglomerated into pressings in, for example, roller presses. Tablets are a typical example of press agglomerates and the particles adhere to each other as a result of van der Waals forces.

5. Sintering

In sintering, highly concentrated suspensions are subjected to high temperatures in an oven and the surface of the particles is partially melted. Bridges formed out of solids are created at the particles' points of contact when the suspension is cooled down.

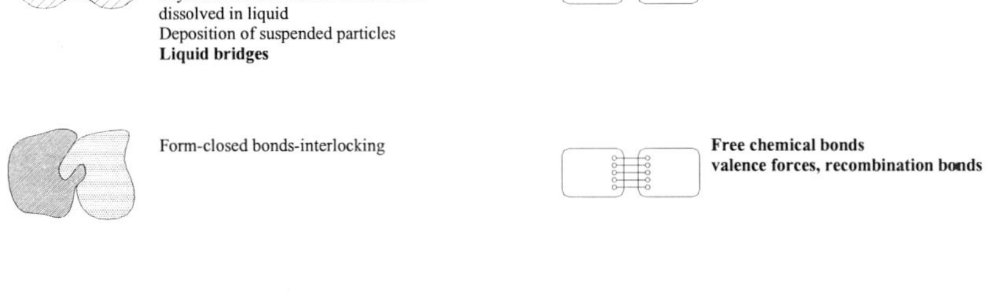

Fig. 9.4 Classification of binding mechanism of agglomeration following Pietsch [95]

Besides the classical pelletising discs, growth agglomeration in particular is performed in solids mixers. [79]. The growth of agglomerates is produced by the mixing and rolling action (Froude number less than 1) or through turbulently "stirring" suspensions (Froude number >>1), bringing about particle collisions. If when a collision occurs the forces of attraction are greater than those of separation operating on the product the fine particles remain attached to each other (by coalescence). If additional primary particles rest on these agglomerates in the form of layers the latter increase in size (through the snowball effect). As opposed to agglomeration on pelletising discs the rolling action in a mixer is caused by the mixer's tools or pneumatically in the case of fluidised beds. If agglomerating liquid is sprayed onto the mix this either produces hard agglomerates or low density instant products according to the stress placed on the product by the machine's tools. The rolling process induced in a horizontal mixer by the motion of its impellers places less stress on the particles and above all is more uniform than that produced in large pelletising discs. The product is consequently more porous. However, the agglomeration strength is reduced to. As higher impact forces are stressing the particles during the granulation, as higher is the strength of the granules. The agglomerates' robustness is however also greatly determined by subsequent drying or sintering processes.

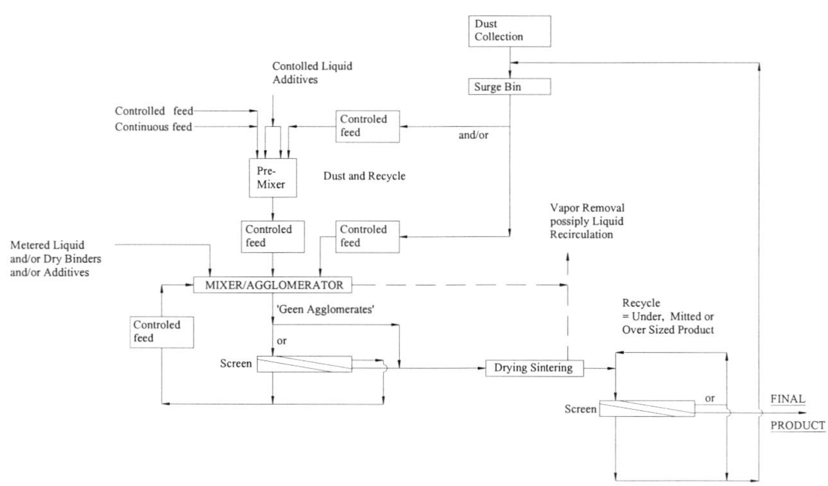

Fig. 9.5 Continuous Granulation following Pietsch [98]

The kinetics of agglomeration are of interest since in most continuous systems they determine the residence time and thereby the system's size. Koch and Sommer [71] have produced a model of the kinetics in a continuous fluidised bed granulator and compared the results with the findings from experiments. F. Hoornaert et alii have ascertained the kinetics in a 50 litre plough share mixer.

Mixers with a high circumferential speed are widely used in the pharmaceuticals sector [97]. Here the shear forces are so great that the snowball effect does not arise and coarse particles and agglomerates grow by coalescence. The proportion of liquid is higher (wet granulation) and the result is granulates with a high density.

The Continuous Granulation Process
Pietsch [95, 98] gives a general overview of the design of agglomeration processes. The overall continuous granulation processes is illustrated by Fig. 9.5 (following W. Pietsch). The particle size granulation of the feed influences strongly the granulation. Before the agglomeration itself the individual components have to be fed and premixed. The product leaving the continuous granualator (mixer) will show a particle size distribution. Unchanged feed particles have to be separated (e.g. by sifting) and recirculated into the process. Oversized particles are ground and conveyed to the feeders. The granules in the desired size range are further processed by drying or sintering.

9.3 Heat transfer

Heat from solids can be dissipated (cooling) or transferred (heating, drying etc.) very well in solids mixers as a result of the material's intensive movement. If heat reaches the solids via special heating or cooling surfaces this is termed *indirect* heat transference. The coefficient of heat transfer in an agitated, thoroughly mixed layer of solids is significantly higher as opposed to that with a static layer of particles since those on the mixer's wall are constantly replaced. Wunschmann measured heat transfer for the solids being in movement or static [100]. For heat transfer a mixer is mostly fitted with a jacket in which flows the heating medium. Hollow shafts or paddles are used less frequently. Please refer to the relevant chapter in the VDI Atlas of Heating [80] and detailed literature [101] to calculate heat transfer coefficients.
In *direct heat transference* the solid is in contact with a hotter (or colder) medium (in most cases a fluid). The cooling of solids in a mixer through the addition of liquid nitrogen is mentioned by way of an example.

Liquid is thermally removed from moist solids through *drying*. To do this the liquid has to be converted into steam and educted from the solids. A distinction is drawn between convection and contact drying. In *convection drying* the heat required for evaporation is transferred by a desiccant taking the form of gas or steam. In *contact drying* the necessary heat is piped in whereby the material to be dried is fed over or rests on heated surfaces. Other methods are radiation drying, high frequency drying and freeze drying. *Vacuum drying* is used for highly temperature sensitive products. The pressure is lowered so that the water is evaporated off at considerably lower temperatures.

In this chapter it has only been possible to touch on the many processes performed in solids mixers. Solid phase reactions are also possible when combined with the above methods. Lastly, this shows once again that mixing is an essential step on the path from educt to product.

10 Designing Solids Mixing Processes

10.1 Goal and Task Formulation

An essential prerequisite for the efficient design of a mixing process is a clear, exact and comprehensive formulation of the task and objective. Applying Table 10.1 as a check list guarantees a systematic formulation of the *mixing task* along with the major *formative conditions*. Priority *objectives* covering the economic requirements, quality targets and operating conditions have to be met when engineering a mixing system. Besides a definition of the stipulated quality of the mix and an average production throughput (minimum or maximum) the *quality target* can also include additional physical (moisture, grain size, temperature) and chemical properties which the mixed product is to have. Furthermore, the general principles of quality assurance frequently demand production documentation. This means that material batches must be coded, mixture recipes recorded and the flow of materials in and out balanced out against their inventories and consumption.

Not to be wondered at is the fact that *formative economic conditions* such as, for example, the available sums for investment, the way the investment is planned as also the use of existing space essentially determine the actual technical features of a design when it is put into practice. Specifications arising from the mixing system's operation are grouped together under *formative operating conditions*. These set the requirements on, for example:

- staff numbers and training
- process monitoring, process management system design and the degree of automation
- operating, cleaning, maintenance
- safety/dust/explosion/emission protection and the alarm system.

Table 10.1 Check list for formulating a mixing task

A) Mix recipes (mixture composition)
- Number and designation of the recipes
- The preparation's composition (the ingredients' percentages and margins of accuracy to be observed, particularly in the case of low dosage ingredients)
- The percentage of each recipe as part of the total production output
- The frequency with which the recipe is changed and any desired sequence
- Cleaning operations when a recipe is changed (dry, wet, cleaning in place CIP)
- Sampling and analyses

B) Ingredients
- Designation
- Origin, supplier, packaging
- Bulk density, solids density
- Grain size (grain size distribution) and shape
- Flow properties, gradient
- Abrasiveness
- Moistness (damp, hygroscopic, dry)
- Temperature, sensitivity to thermal stress
- Sensitivity to mechanical stress (crushing, abrasion, fracture)

C) Product (mixture)
- Mix quality
- Bulk density
- Fluidisability (air take-up during mixing)
- Tendency to segregation
- The mix's flow properties
- Agglomeration, disagglomeration required

D) The mixer performance
- Mix performance: production volume per unit of production (average, minimum, maximum)

For batch mixers:
- Batch mix size (final volume after mixing)
- Start-up filling level
- The filled mixer's idle time

For continuous mixers:
- The production volume with an unchanged recipe
- Feed/mix output tolerance range

E) Integrating the mixers into the system
- Material flow diagram (average, maximum and minimum figures)
- The ingredients' in and outflow
- Spatial requirements, height, layout
- The mixture's usage
- Storing, feeding and weighing devices
- The type of process inspection, process control, storage and data exchange
- Safety requirements

F) Mixer design
- Raw material, surfaces and the in and outflow configuration
- Heating, cooling, inertising, pressurisation, vacuum
- The addition of liquid into the mixer
- Disagglomeration
- Current, steam and water connections, adjutants, types of protection, protection against explosion

G) Formative economic conditions
- Investment costs
- Maintenance, running and staff costs
- Profitability

10.2 The Choice: Mixing with Batch or Continuous Mixers

The advantages of continuous and discontinuous mixing process were discussed at the beginning of Chapter 8. Continuous mixers are particularly beneficial where there is a risk of the mixture separating, which especially occurs whilst the prepared mix is being handled or poured into holding bins since continuous systems avoid the need for such intermediate storage. As a result of its compact dimensions a continuous mixer can be installed facing the next processing stage with greater ease and the produced mix fed directly into the further processing operations. In a continuous process only a small quantity of material is stored, which considerably simplifies safety designs in the case of dangerous products or where there is explosive dust in the atmosphere. Table 10.2 gives a comparison of discontinuous and continuous mixing processes. Besides the number of ingredients in the mix a decisive feature in selecting the process is the individual component's flow volumes. Since the feed's constancy can only be maintained with a limited degree of accuracy at continuous feeding rates below 300 g/h, ingredients with low flow volumes necessitate a premixing operation.

Table 10.2 Comparison of discontinuous and continuous mixing processes

Implementation data	Discontinuous	Continuous
The number of ingredients	As many as wanted	2-10. Any more ingredients are usually combined in a premix.
Frequency with which the recipe is changed	Several times per hour	A recipe must remain unchanged for several hours
Cleaning frequency or idle time	Several times a day	Once a day or less
Production output, throughput	Any rate	More than 100 kg/h. Exception: feeding laboratory extrusions
Risk of separation	Present, therefore there must be short transportation paths, few intermediate silos	Low risk when the material is taken directly to the next processing stage or directly drawn off
Spatial requirement	Large amount of space and intermediate silos required for machines with a throughput greater than 5000 kg/h	Low spatial requirement even for machines with a high throughput
Requirements placed on the equipment	Simple feeding but high demands on the mixer	Accurate continuous feeding (feeding scales necessary) but low demands on the mixer
Safety	Steps have to be taken in the case of materials with a risk of explosion	The small quantities of material present during processing have a low potential risk, which simplifies safety design
Automation	Variable degree of automation	Contained in the processing

10.3 Feeding and Weighing Equipment for a Batch Mixing Process

The number of mix cycles multiplied by the usable mixer capacity gives the set mixture output per hour. The mix cycle consists of (Fig.10.1) the filling, mixing, discharge and idle times. To this is added in special cases the time taken for sampling and analysis and that for associated processes such as disagglomeration and granulation.

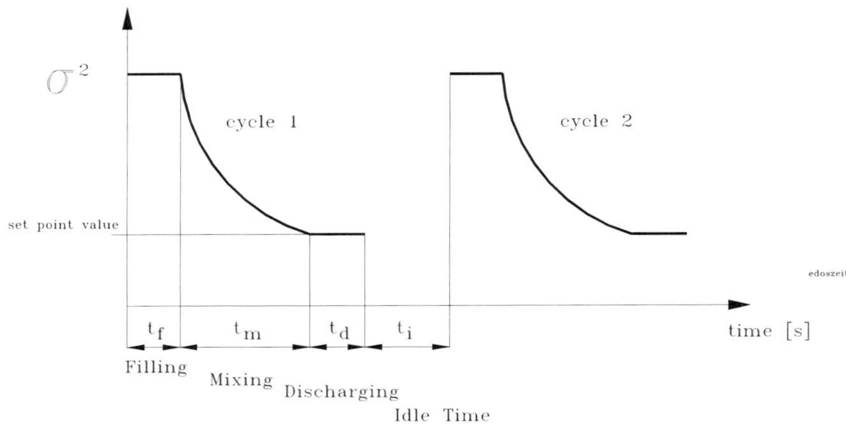

Fig. 10.1 Cycle times when batch mixing solids

The capacity (throughput rate) of a batch mixing process having a mixture charge with a mass M gives the following as shown in equation 10.1:

$$\dot{m} = \frac{M}{t_f + t_m + t_d + t_i} \left[\frac{kg}{s}\right] \tag{10.1}$$

The mixing time t_m depends on the selected mixer design and size, the filling time t_f on the system's configuration whilst the discharge time t_d depends on both the mixer's design and the system's layout.

The choice of feed and weighing devices is determined by the number of ingredients, their mass and proportions, the throughput volume, the stocking and mode of delivery, the spatial circumstances, degree of automation etc. In the simplest case the ingredients are manually weighed into the mixer. Where there are higher requirements in respect of accuracy, safety and recording, a hopper scale represents a simple device for weighing and releasing the components into the mixing equipment (Fig. 10.2). The measured section comprises the weighing hopper with or without a cover and with an outlet device. The weighing units are mechanical or electromechanical and in both cases frequently a hybrid construction.

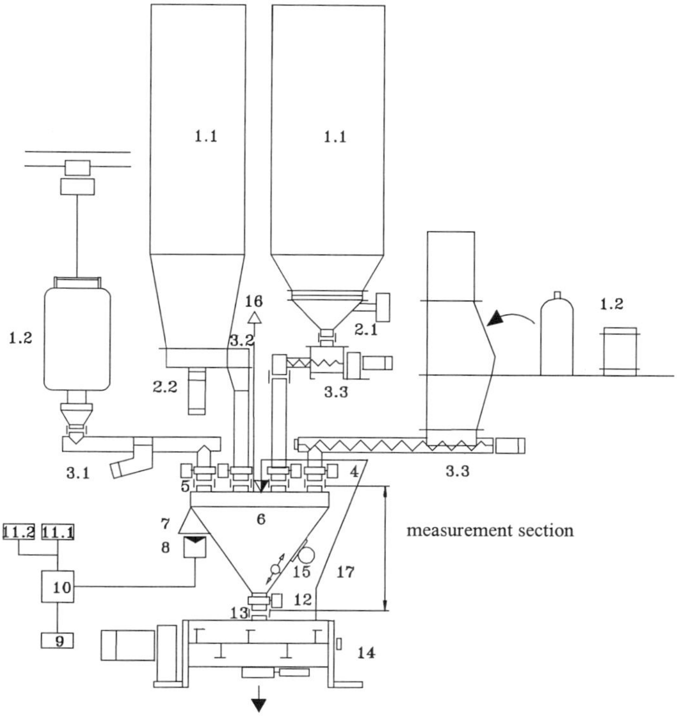

Fig. 10.2 Weighing hopper with additive weighing for feeding a batch mixer
1.1 storage silos; 1.2 big bag, bag, drum; 2.1-2.2 dischargers; 3.1-3.3 feeder units; 4 cut-off; 5 flexible connections; 6 weighing hopper; 7 support for gravity force; 8 gravity operated sensor (load cell); 9 set point; 10 weighing analysis and regulation; 11.1 measured value indicator or output; 11.2 recorder (printer); 12 cut-off; 13 flexible connection; 14 mixer; 15 discharger; 16 dust extraction and weighing hopper ventilation; 17 mixer ventilation

The feeding and weighing process is performed as an *additive* (positive) fill weighing (Fig. 10.3). Since in most cases several ingredients are fed into the same hopper only a gross weighing operation is possible. Any discharging error is recorded (by net weighing) through logging the difference in weight during emptying. Because of the differing ingredient volumes it is often desirable to change the measurement range (as can be done in multi range or composite scales). A second or more weighing hoppers with lower weighing ranges are frequently engaged, which gains time and produces greater accuracy with minor ingredients.

Negative net feeding (subtraction weighing) is used (Fig. 10.3) for small quantities of individual components or cohesive, sticky solids to avoid errors caused by the varying quantities of them adhering to the weighing hopper after it has been emptied. This type of weighing only records the quantity of material which has actually been metered out and weighed. Loss-in-weight (differential) feeders are also used on occasion in batch mixing.

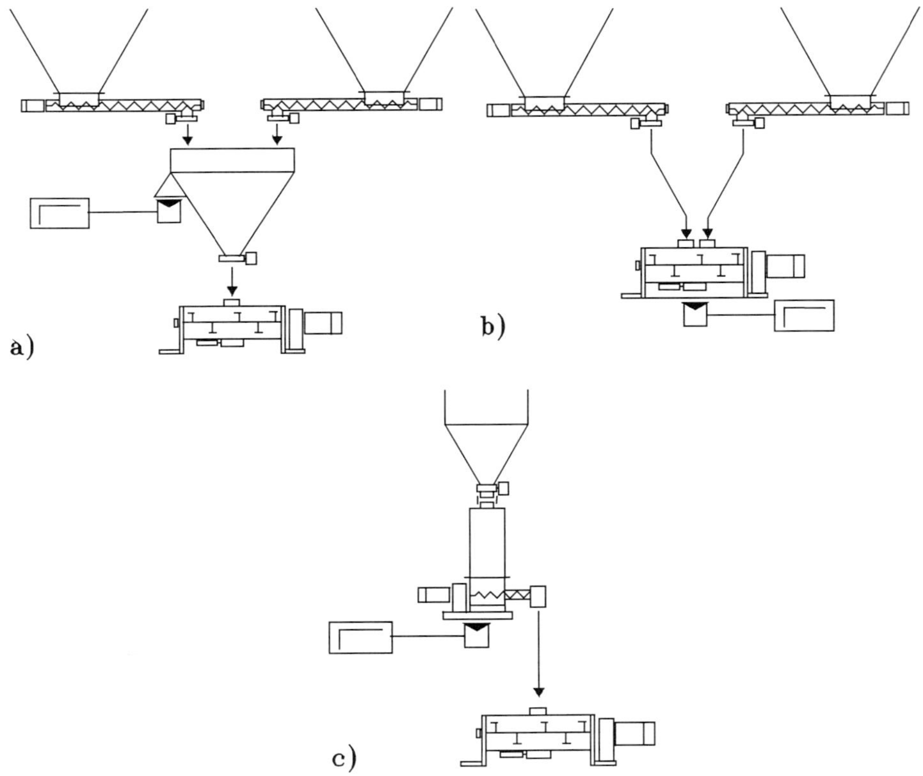

Fig. 10.3 **a)** Positive net feeding into the weighing hopper; **b)** positive gross feeding of all the ingredients into the mixer; **c)** negative net feeding (subtraction feeding of an individual ingredient)

Instead of stationary weighing hoppers, sequence weighing can be performed using travelling scales in mixer systems with a large number of ingredient silos (Fig. 10.4).

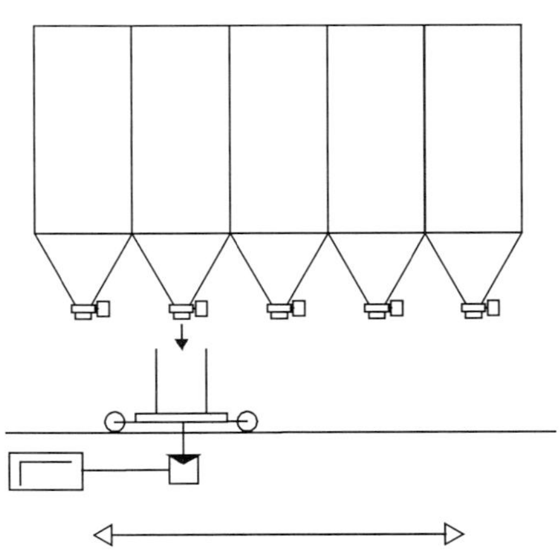

Fig. 10.4 Sequence weighing

The weighing operation in the mixer is generally performed by a multi purpose weighing hopper (Fig. 10.5).

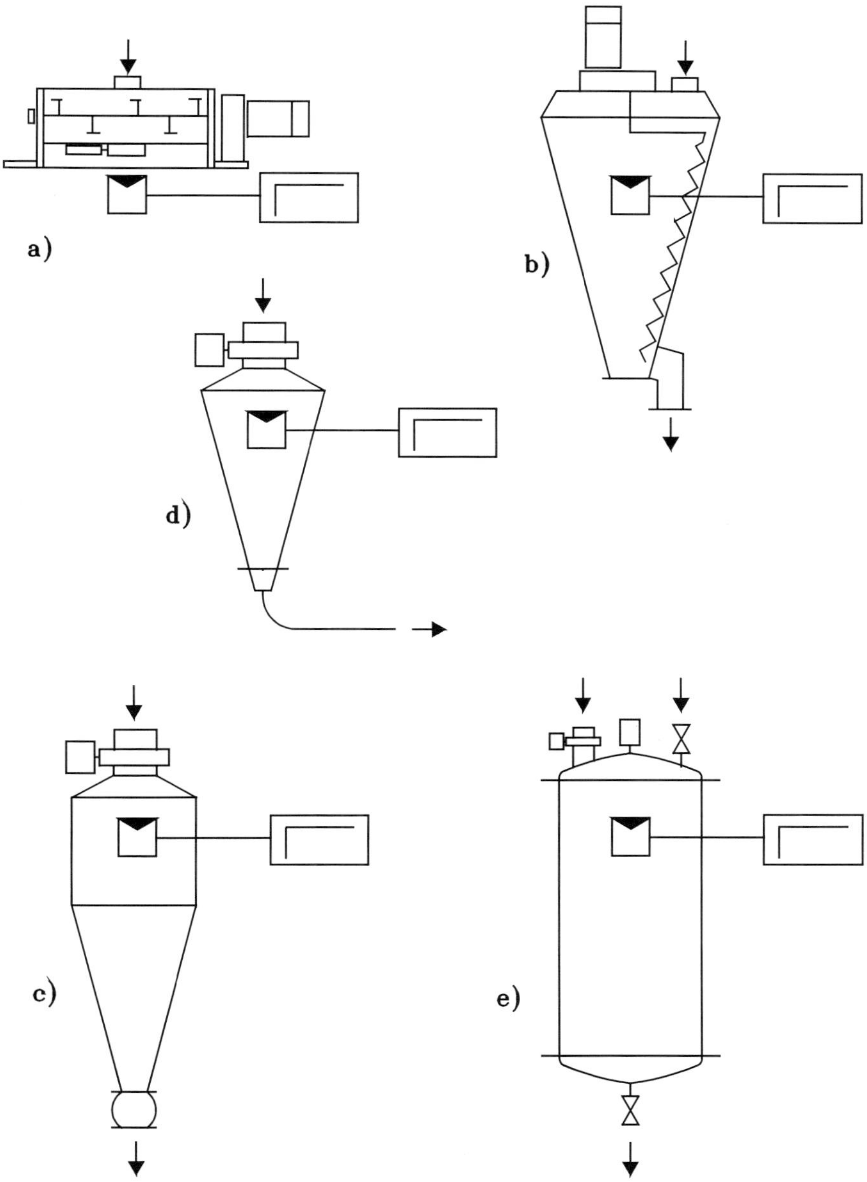

Fig. 10.5 Multipurpose weighing hoppers: a) horizontal mixers; b) vertical mixers; c) pneumatic air mixers; d) pneumatic pressure conveying vessel; e) reaction vessel

The combination of weighing with pneumatic feeding the mixer can be performed by pressure (Fig. 10.6) or vacuum system. A pressure driven system allows dense

Vacuum conveying is useful for a large number of ingredients, as in the case of the solution shown in Fig. 10.6c. It works only if the system can be started with the pipework filled with material.

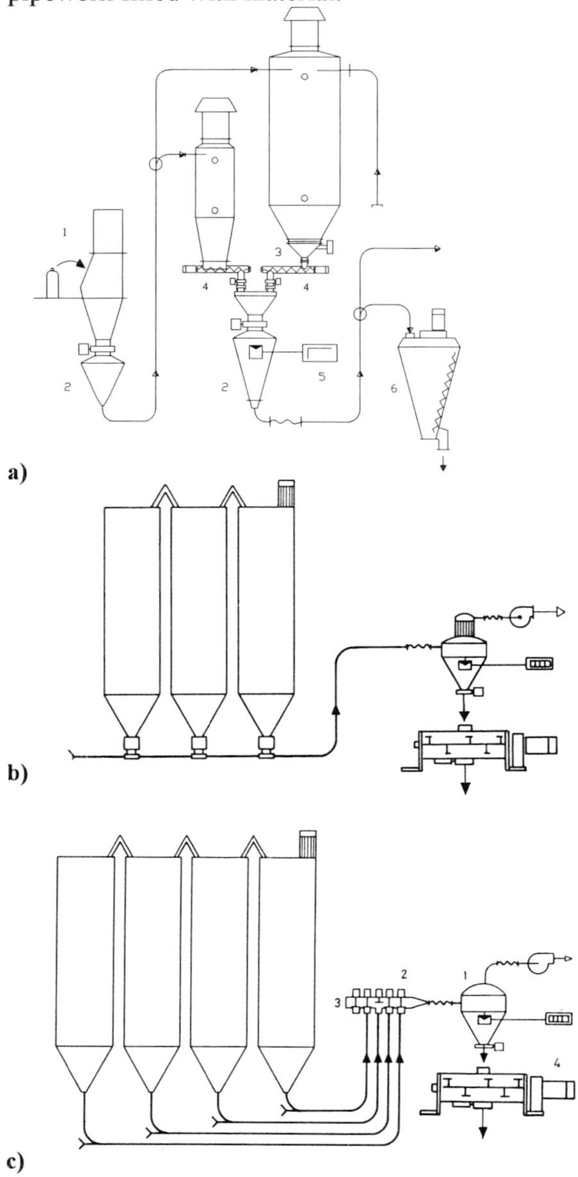

Fig. 10.6 Combined weighing and pneumatic filling of a batch mixer
a) Feeding the mixer by pneumatic pressure conveying: 1 filling; 2 charging the silo; 3 discharging; 4 feeding; 5 scales; 6 mixer **b)** Vacuum conveying with feeding at pipline inlet **c)** Vacuum conveying with feed valve at weighing hopper: 1 weighing hopper; 2 valve switch; 3 infiltrated air; 4 mixer

Hopper and mixer ventilation and dust extraction

During filling (feeding), the air in the weighing hopper and the mixer is expelled and must be able to re-enter them during discharging, otherwise flexible connections will be torn or the emptying process hindered. The discharging hopper must also be ventilated accordingly. To avoid weighing errors, ventilation and thereby dust extraction must be performed in such a way that fluctuating underpressures in the vacuum filter do not exercise any downwards acting force on the weighing system.

Guidelines on the size of the weighing hopper and the weight scale to be used

As a rough rule of thumb it can be said that the smallest ingredients should weigh 2 to 5% of the nominal load in the case of electromagnetic or electronic scales and 10% for mechanical scales. You can switch down to scales as low as 1:5 in the case of multi-range weighers and 1:10 or less using the finest graduations on a composite unit.

Feeders

Screw feeders, vibrating tubes, rotating, bin discharge devices, rotary valves, discharging belts as well as feed valves for free fall feeding come under consideration. The latter, used both as sealing and fast shut off devices, must be able to close whilst immersed in the product's flow. Besides suitability to the product, a wide range of settings and fine flow uniformity are also important. Three phase motors are pole changeable up to 10:1, whilst the pole changing on DC motors can be set up to 50:1, frequency controllers up to 100:1, step motors up to 100:1, and electromagnetic vibrating troughs up to 30:1. Attention should be paid to the continuous load which can be placed on the gear drive over the whole range of settings.

Calculating the feeding and weighing output of a weigh hopper as part of the mixing system's performance

The batch mixer's throughput is determined by the cycle time (filling + mixing + emptying + idle time) as shown in equation 10.1. In the case of hopper scales during the mixing and mixer discharge the ingredients can be feed and weighed. The feeding process (Fig. 10.8) often forms a temporal bottleneck in respect of each of the ingredients. Simplified calculations can be performed to give a coarse approximation of the feeding and weighing time. The total cicle time is:

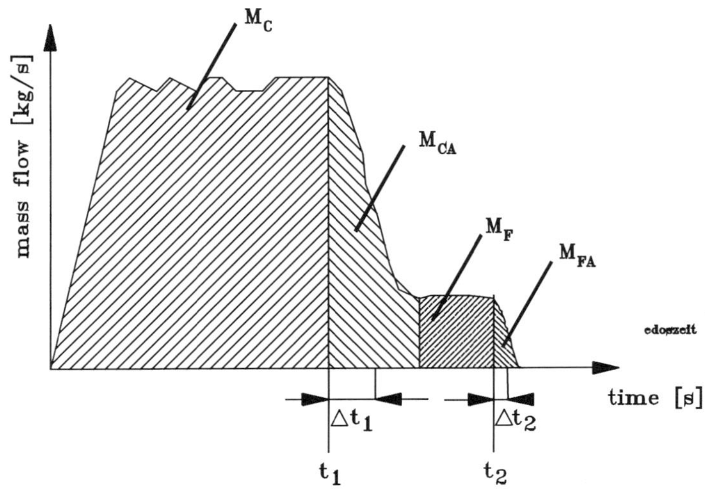

Fig. 10.7 The cycles in a discontinuous feeding system: $\mathbf{M_c}$ coarse feed mass volume; $\mathbf{M_{CA}}$ coarse feed after-run; $\mathbf{M_F}$ fine feed mass volume; $\mathbf{M_{FA}}$ fine feed after-run

Coarse stream feeding time + 6 seconds for fine stream feeding for each ingredient + 10 seconds for opening and closing the weighing hopper discharge valve until starting the next recipe + the time for emptying the weighing hopper. A detailed layout of feeding processes is given by Gericke [36, 81].

10.4 System Errors in a Batch Mixing Installation

Deviations of the physical (actual) value from the set point value of an ingredient's percentage in a sample taken from the filled package unit are designated as system errors:

$$f = x - x_{sp} \tag{10.2}$$

In addition to the unavoidable concentration fluctuations in solids mixtures (cf. Chap. 3) system errors are produced by a multitude of additional causes. Care must also of course be taken against separation of the product's constituents until that product reaches the point and time of its usage. Care in choosing the section to be observed can be very helpful in recording and eliminating these system errors. Fig 10.8 shows the section to be observed for ascertaining errors in a batch mixing system. System errors are caused by the process, the feeder unit, the equipment employed, the material to be blended, the environmental factors and storage. The system error's absolute value is in turn related to the size of the sample.

The system error is estimated by the sampling method:

$$\bar{f} = \frac{1}{n}\sum_{i=1}^{n}(x_i - x_{sp}) = \frac{1}{n}\sum_{i=1}^{n} f_i \tag{10.3}$$

Table 10.3 distinguishes system errors which can arise before, during or after the mixing process.

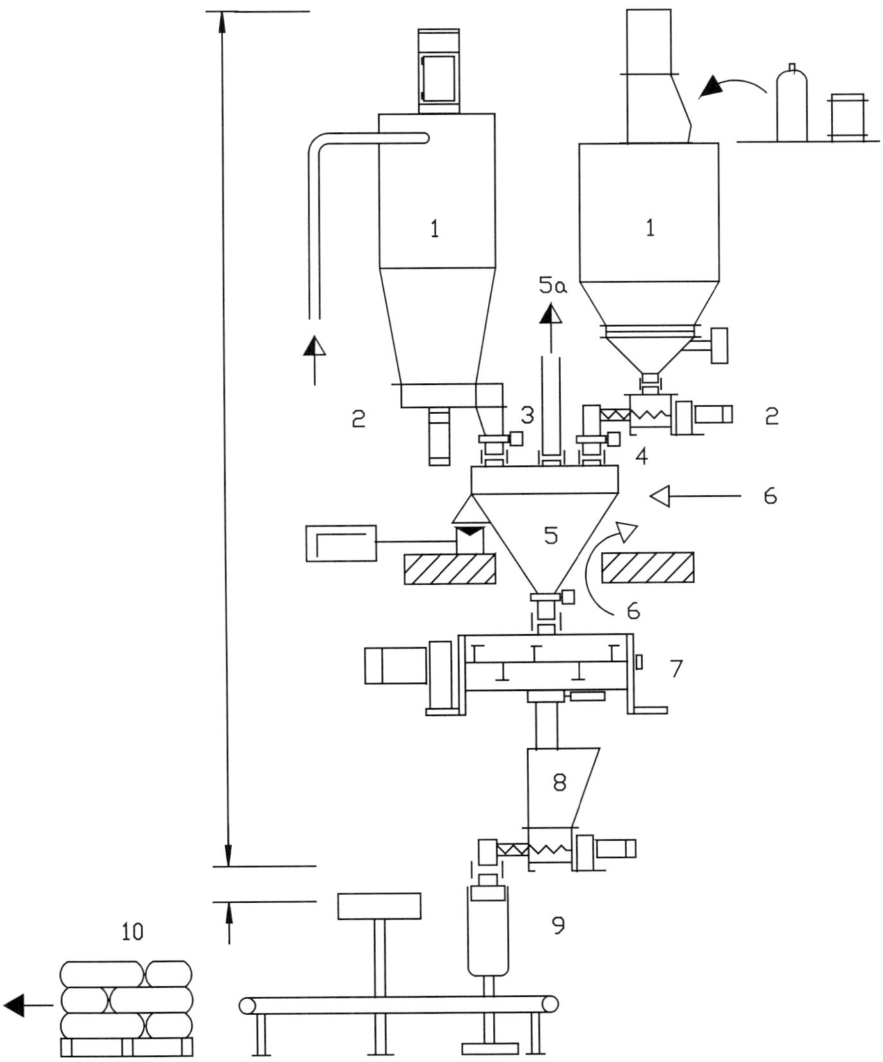

Fig. 10.8 The section under observation for ascertaining the system error in a batch mixing system: 1 storage silos with product feed; 2 discharge and feeding units; 3 discharge devices; 4 flexible connections; 5 weighing hopper with 5a ventilation; 6 wind, air draft; 7 mixer; 8 mixing discharge hopper; 9 drum filling; 10 drum or bag storage and transportation

Table 10.3 Causes of system errors in a batch mixing process

Errors prior to the mixing process
a) Errors in the feeding and weighing unit
- Feeding errors, systematic and random deviations attributable to the feeding system
- Random and systematic measuring errors in the form of a deviation of the indicated or output value from the true (actual) one
- Errors whilst emptying the weighing hopper and which are caused by various residues in the hopper and delivery unit from the weighing operation
- Conveying errors caused by residues in the flexible connections, connection pipes of weigh hopper and mixer
- Dust extraction and ventilation malfunctions
b) Errors caused by external sources
- Influences exerted by flexible connections on lines to and from weighing hoppers and mixers
- Mechanical influences such as oscillation, contact, electrical influences and electromagnetic interference fields
- Wind pressure on the weigh hopper, drafts
- Temperature and the effects of variable pressure
- Operating errors, e.g. when inputting the set point, when weighing manually and when making up a report
c) Errors in preparing the product
- Not removing foreign bodies (remedy – control sifting)
- Agglomerations in bags or silos (remedy – disagglomeration e.g. by nibbler)
Errors in the mixing process
- The mixer does not attain the required mix quality (remedy - lengthening the mixing time, a finer grain structure in the input ingredients, perhaps a different type of mixer and the addition of liquid in the event of separation)
- Overmixing: the product suffers damage during the mixing process through mechanical thermal effort or heat
- The undesired formation of agglomerates when liquid is added
Post mixing process errors
- Separation during discharging, in the discharge hopper or in the filling drum or bag (remedy - design the discharge hopper for mass flow and equip it with the appropriate outlet device) If the tendency to separation cannot be overcome you should investigate changing to a continuous mixing process
- Residues in the mixer or the discharge hopper when the recipe is changed (cross-contamination)

10.5 Feeding and Weighing Equipment for a Continuous Mixing Process

As shown in Chapter 8 the mixing process is essentially influenced by the quality of the continuous feed. The objective is to feed set quantities of material per unit of time in an uninterrupted product flow. Feeding can be performed **volumetrically** by feeder equipment which draws material by volume (screw feeder, vibrating feeders, rotary valves, belt/agitator/disc or ring groove feeders) or **gravimetrically** by controlled feeding using weight or mass as the control value (loss-in-weight (differential) feeder, belt/rotating/screw weigh feeder scales).

A continuous feeding system consists of:

- *the feeder unit*: a distributor device and adjuster for producing the volumetric flow of the material to be fed, continuously and with the highest possible degree of volumetric feeding accuracy.
- *the measurement section*: which takes volumetric measurements using the geometry and motion of the feeder, or gravimetric measurements by recording

the mass or weight of the material flowing through or delivered.
- *the control system*: which sets the reference values by controlling or regulating the feeder drive.

There are three versions of a continuous, multi ingredient feeding system:
1. Each ingredient is autonomously adjusted, independently of the others.
2. Each ingredient is adjusted to a specified ratio of the totality of the ingredients - *recipe feeding* (Fig. 10.9). Recipe proportions by weight, are entered as absolute values or percentages into the main recipe or system controller for a given total flow m. The percentage proportions remain unchanged when the total feeding volume is adjusted.

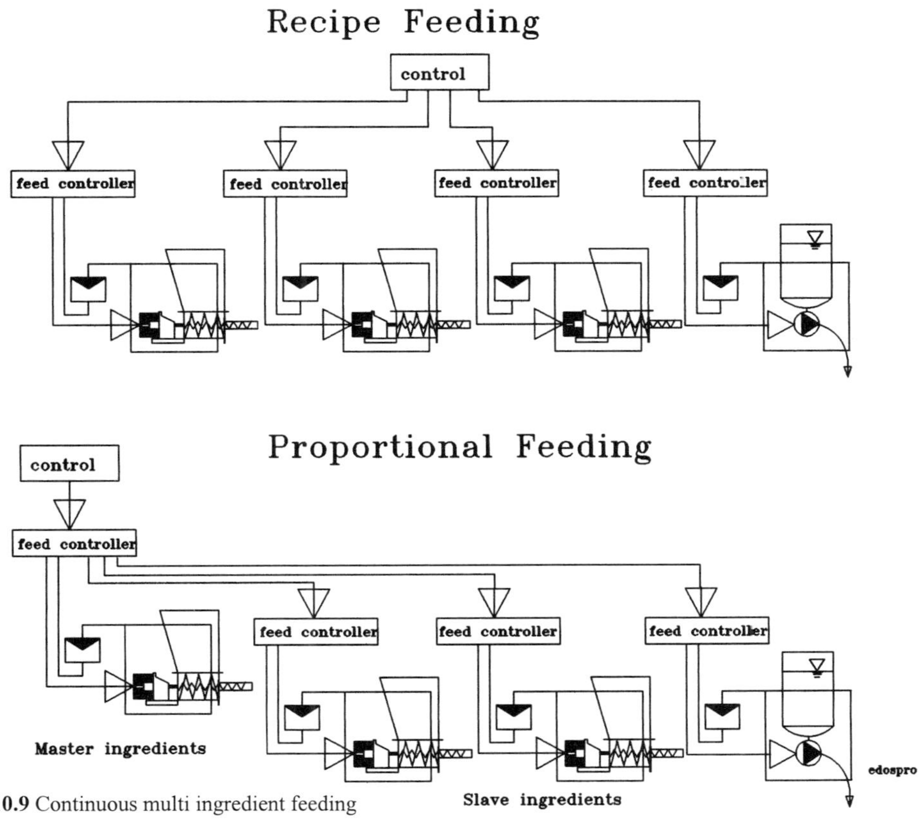

Fig. 10.9 Continuous multi ingredient feeding

3. One ingredient serves as the primary constituent or master and the other (slave) ingredients are fed in relation to it - *proportional feeding* (Fig. 10.9). The master ingredient is fed at a chosen constant rate. If this stream is altered the proportions of the master ingredient are retained. In som systems the master ingredient can be a variable ("wild flow") feed flow which is continuously measured by a set of test scales and serves as the guide value for the rest of the ingredients.

Free flowing and uniform solids (granulates) can be fed *volumetrically*, as can liquids. All other solids as well as minor ingredients in particular (where the flow is only a few kg/h) must be fed *gravimetrically*. Gravimetric feeding offers greater accuracy even over periods of hours or days (long term constancy) and is also suited to the feeding of materials with a fluctuating bulk density or flow properties, such as cohesive powders and liquid additives with a variable viscosity or a varying quantity of gaseous inclusions. Gravimetric feeding also enables the actual metered volume to be reported back for the purposes of recording, taking printouts and storing, as well as data transference to the process control, information management and alarm systems.

The cost of investing in a gravimetric feeding system for one ingredient is a factor of 3-4 times higher for a feeder and control unit than for an unregulated volumetric feeding system. Maintenance and servicing costs are likewise higher but this depends greatly on the system you choose.

Gravimetric continuous feeding

Fig. 10.10 shows the control circuit for continuous feeder belt scales. Belt scales are predominantly only used for uniform, dust free materials or for very fragile materials.

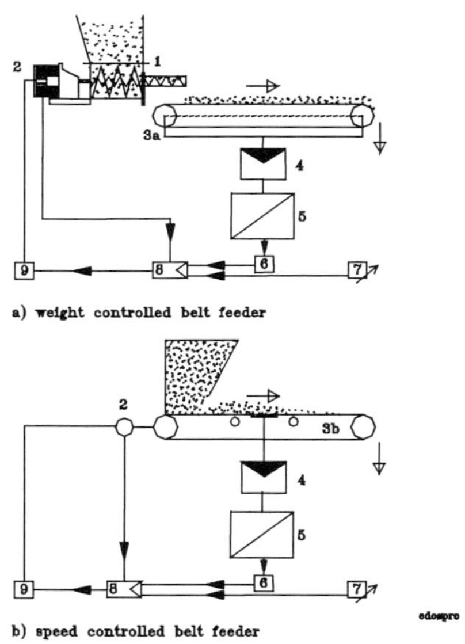

Fig. 10.10 Weigh belt feeders for the gravimetric, continuous feeding of solids
Control elements: 1 feeder unit; 2 drive gear. Control section: 3a belt scales; 3b weighing section; 4 load cell; 5 controller and 6 signal processor; 7 set point input; 8 set point/actual value comparison; 9 motor controller

Detailed descriptions are given by Gericke [8], Vetter [82] and Hlavica [83].*Loss-in-weight feeders* (=differential weigh feeders) can be used in all cases where their operating principle is as shown in Fig. 10.11. The feeder unit (1) with its built-on hopper (2) is supported on or suspended over the weighing system (3). The feeder unit and hopper are fitted with flexible connection collars at their in- and outflow points. The refill unit (4), in most cases fitted with a cut-off valve (5), refills the feeder unit to its upper limit m_{max} or G_{max}. After the refill unit has been switched off and the cut-off device closed, the load cell (6) measures the weight or mass volume at regular intervals (counted in fractions of a second). A comparator (7) calculates the weight loss per unit of time dG/dt or dm/dt which corresponds to the physical (actual) feed rate. After comparing the actual value in the feeder controller (8) with that of the set point entered by the controller (9), the feeder unit's drive speed (11) is adjusted by the motor controller (10) so that the feed flow matches the selected set point. The tachometer (12) reports the rpm to the motor controller in order to correct the motor speed precisely (via a sublayered control circuit). As soon as the minimum filling level m_{min} or G_{min} has been reached the drive is switched over to a constant rpm and refilling up via the refill unit is initialised. Gravimetric control recommences after the maximum filling level m_{max}/G_{max} has been attained and the weighing system stabilised. 10 to 20 or more (as required) refills occur per hour at a sufficiently high refill capacity to keep down the unregulated proportion of time.

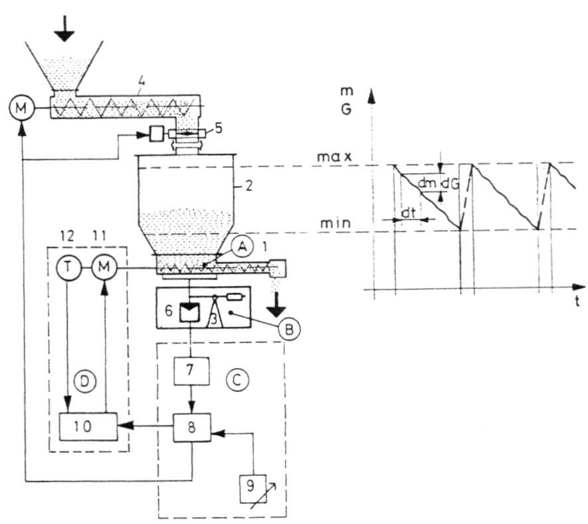

Fig. 10.11 Loss-in-weight feeder (differential weight feeder): **A** feeder unit; **B** weigher; **C** feeder controller and regulator; **D** feeder unit drive gear; 1 feeder unit; 2 built-on hopper; 3 weighing system; 4 refill unit; 5 discharge valve; 6 load cell; 7 comparator; 8 feeder controller; 9 set point regulator; 10 motor controller; 11 feeder unit drive; 12 tachometer

Belt scales are generally used in the case of a proportional feeding system (slave feeding) where the main ingredients' throughflow volume is measured. Coriolis force throughflow measuring equipment comes under consideration for higher outputs (over 1 t/h and for non-cohesive materials).

Feed and accuracy (feeding errors)

The continuous feed's accuracy is of great significance to the system accuracy of a mixing installation. In order to characterise the feeding's accuracy the feed rate range of solid ingredients has to be given along with an average bulk density. Taking a basis (cf. Fig. 10.12) of n feed flow samples of $m_i = M_i/\Delta t$ in a periodΔt (recommended values are n = 30 seconds and Δt = 60 seconds, but this can shorter according to the case in question) the following can be established [99]:

- **The feed constancy** as a short term scatter of the feed flow's values around their mean point

- **The feed accuracy** or **feed error** as a deviation of the average values from the reference value (set point)

Fig. 10.12 illustrates the procedure for establishing the actual value $m°_{act}$. The mass flow $m°_i$ is ascertained at specifiable time intervals Δt (for example 60 seconds) and the feed constancy calculated from a series of n measurements.

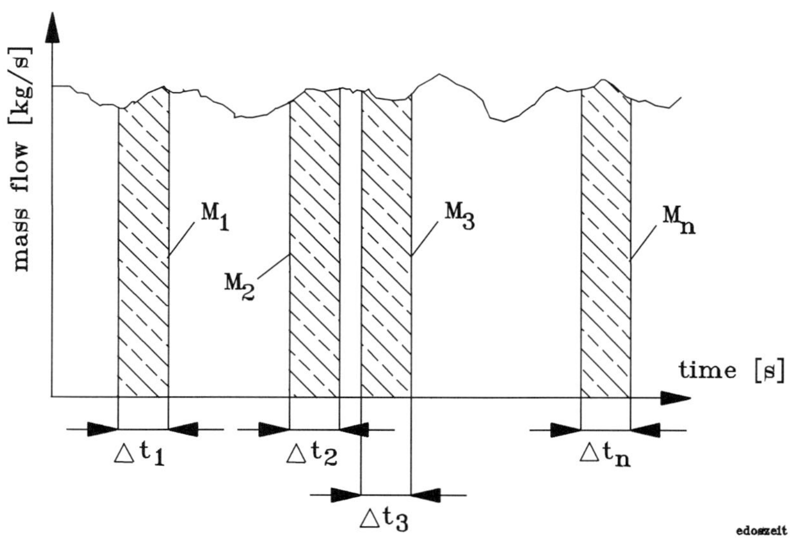

Fig. 10.12 Establishing the feed flow $m°_{act}$

The average deviation or the empirical standard deviation are used to define the feeding accuracy:

average deviation $$AD = \overline{\dot{m}} - \dot{m}_{sp} = \left(\frac{1}{n}\sum_{i=1}^{n}\dot{m}_i\right) - \dot{m}_{sp} \qquad (10.4)$$

empirical standard deviation $S = \sqrt{\frac{1}{n-1}\sum_{i=1}^{n}\left(\dot{m}_i - \overline{\dot{m}}\right)^2}$ (10.5)

Subject to the prerequisites of the mass flow having a standard distribution and there being sufficient feed flow sampling, the following percentages of the actual mass flow values in the interval range lie around that of the average:

68.3% within ± 1 s

95.4% within ± 2 s

99.7% within ± 3 s

The *relative feed constancy* is defined in equation 10.6:

$$S_{rel} = \frac{|S|}{\overline{\dot{m}}} \cdot 100\,[\%] \tag{10.6}$$

and greatly depends on the size of the feed flow. Clearly if the recipes and that is to say the feed flows for an ingredient are changed during continuous mixing the feed constancy must be related to one particular feed flow.
The *feed accuracy* or *feed error* is the average deviation, i.e. the average flow's deviation from the set point at a specified point in time:

absolute: $$\Delta\overline{\dot{m}} = \overline{\dot{m}} - \dot{m}_{sp} \left[\frac{kg}{h}\right] \tag{10.7}$$

relative: $$\Delta\overline{\dot{m}}_{rel} = \frac{\overline{\dot{m}} - \dot{m}_{sp}}{\dot{m}_{sp}} \cdot 100\% \; [\%] \tag{10.8}$$

with $$\overline{\dot{m}} = \frac{\sum_{i=1}^{n} \dot{m}_i}{n} \tag{10.9}$$

The feed error and feed constancy is also governed by the test method. Measurement errors, both random and systematic, are to be taken into consideration as time and weight errors. Only where there is a sufficiently precise measurement procedure does the feed error or feed constancy ascertained by experiment describe the actual feeding process.

Fig. 10.13 shows the curve of a feed flow over several days. At the beginning and towards the end it is indeed virtually time constant but the absolute deviation from the set point becomes plainly greater. When a solid's bulk density diminishes over a long period of time is being fed volumetrically the feed flow drops down if the feeder's rpm is not matched to the altered bulk density.

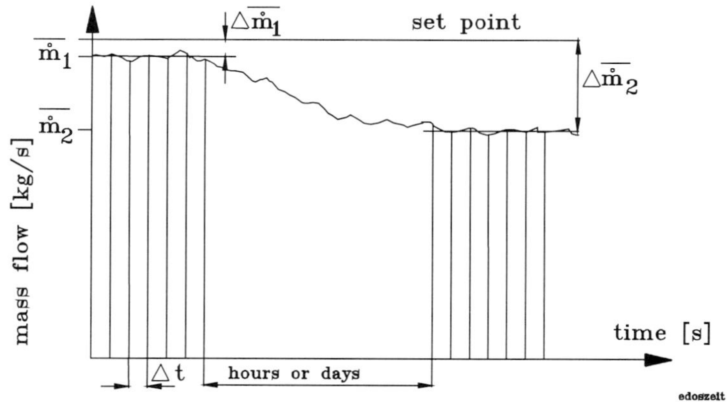

Fig. 10.13 Long term drift in a continuous feeding system

The total feed error in a long term observation is the average value's deviation plus/minus the feed constancy's short term error:

$$S_F = \pm\Delta\overline{\dot{m}} \pm S \tag{10.10}$$

Feed error and final mix quality

The final mix quality found at the outflow from a continuous mixer is composed of the feed and mix errors. According to the way the mixer operates there is a greater or lesser actual mixing effect resulting in a broader or narrower residence time distribution and according to the mixer's capacity, short term feed fluctuations can (taking a mean residence time) to a certain extent be equalised. If we consider Fig. 10.14 it can be seen clearly that a broader short term scatter S is rather more acceptable than a large, long term average value deviation. A continuous mixer cannot exercise any corrective function in the case of the latter.

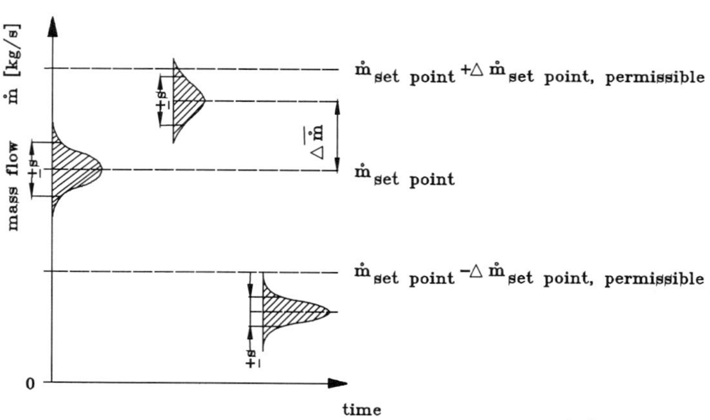

Fig. 10.14 Vetter's [82] description of flow volume distributions and average value deviations at various points in time

10.6 System Errors in a Continuous Mixing Installation

As shown in Fig. 10.15, the section under scrutiny encompasses the ingredients' delivery and discharge, any preparation (e.g. the removal of foreign bodies, breaking up lumps), intermediate storage, feeding, mixing and conveying of the mixture.

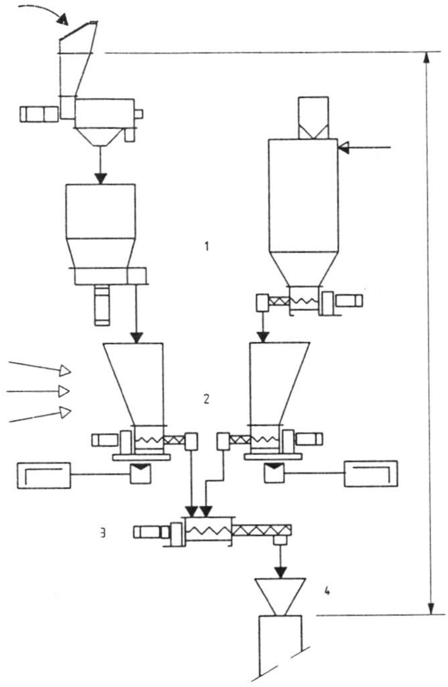

Fig. 10.15 Section under observation for system errors in a continuous mixing system
1 ingredient silos; 2 continuous feeding unit; 3 continuous mixer; 4 discharge hopper, filling machine

The desired feed accuracy influences the continuous mixer's design and size. A small, compact mixer with a low residence time and limited axial mixing requires a high degree of feed accuracy even over long periods of time. In comparison to a batch mixing plant the risk of separation can be better mastered thanks to the continuous flow of material after it has passed through the mixer.

10.7 Protection in Mixer Installations against Explosion and Fire

There is a danger of explosion and fire in mixtures of dusty, flammable solids with air, flammable gas or vapours (e.g. solvent vapours). In general, powders with a particle size of < 500 μm can be flammable if they are distributed uniformly in the atmosphere within a specified concentration range and if the dimensions of the apparatus, i.e the mixer or pre or post mixing hopper are above a minimum size.

The following equipment has to be encompassed by the observed section in the

safety analysis:
1. All silos, hoppers, processing equipment such as mixers, mills, dryers, conveyors, dust extractors etc.
2. All fine powders or dusts in a mix even when being of small percentage by weight but are in the form of fine particles.

A systematic or computer controlled implementation of safety analysis (e.g. using "Dustexpert" [84]) for the processed dusts and across the whole system serves in establishing the protective measures to be adopted. Legal aspects and standards must also be taken in account [85]. The aim is to draw up engineering data on the system to form the basis for its construction and operation.
Safety measures can be *primary*, i.e. directed against the occurrence of explosion or fire or *secondary*, i.e. against the effects thereof within or outside the mixer installation. The ingredients to be mixed, the finished mixtures and the powder obtained from dust extraction can be investigated for their explosive properties [102, 39, 86, 87], in particular the maximum overpressure and the Kst value for the maximum temporal increase in pressure. Values for various particle size distributions of 1900 powders are contained in the handbook on "The combustion and explosion parameters of dusts" [91]. By means of laser ignition, the minimum ignition temperature and energy of single particles are today determined [103].

10.7.1 Protective measures against the occurrence of explosions - primary measures

Protective gas overlay (inertising)
Inertising serves to avoid a potentially explosive atmosphere [88]. Atmospheric oxygen can be replaced by a protective inert gas, mainly nitrogen, so that the reduced O_2 content no longer enables any ignition and combustion. With most dusts a reduction of the oxygen concentration to 12% less 2% for safety, and therefore < 10% by volume, is sufficient. In addition, it is possible to mix powdery solids with an inert or quenching dust, a fine extinguishing powder. Inertising calls for closed systems which are flushed out with inert gas before being put into operation. **Monitoring** by checking the protective gas's overpressure or, as a safer method, by measuring the O_2 content in the dusty section, is performed by explosion-proof equipment.

Avoiding sources of ignition
Besides external sources, ignition is caused by sparks produced mechanically, by heat produced from friction, by static charges, by electrical sparks, by too high a temperature in the material or in parts of the equipment and by chemical reactions. Almost 30% of all recorded explosions in powders are triggered by "mechanical sparks", i.e. from friction and impacts [92]. Steps which can be taken against these are:
- limiting the circumferential velocity of rotating components, in general to less than 1 second, evenutally higher, checked by case test [86],
- constructing one component out of plastic where there is a risk of moving parts coming into contact.

10.7.2 Protective measures against the effects of explosions - secondary measures

An explosion resistant mixer, silo and container design is a possible secondary measure. There are two types of construction, pressure proof or pressure shock resistant [89, 93]. Mixers, hoppers etc. are built as *pressure proof* pressure vessels to a designed and calculated pressure equal to that from the critical dust's explosion and in accordance with the provisions governing pressure vessel construction in force in that country. Where the operating pressure is below the range of a pressure vessel, mixers and containers are constructed as *pressure shock resistant*. The same calculation principles apply. However, deformation is in this case permissible without the mixer ripping open or parts flying off. The calculated pressure is the maximum pressure from the explosion of a combustible material (the coefficient of safety versus the stretch limit, or 2% of the expansion limit for austenite steel, is set at 1).

Explosion pressure release

The maximum explosion pressure can be decreased to a reduced one P_{red} (Fig. 10.16) by opening an orifice with a specified diameter at the right time. The lower the opening pressure P_{stat} the lower the reduced maximum explosion pressure P_{red} to which the mixer's resistance must be designed. For example, P_{stat} = 0.3 bar, P_{red} = 2.5 bar at an unrelieved P_{max} of 8.3 bar. Non reusable rupture discs or explosion flaps open at a burst pressure related to the medium being processed, the temperature and the cross section of the flow. Straight pressure relief pipes are led out into the open air or into special receivers in order to protect personnel in the area. An automatic burst indicator as well as proper maintenance with regard to contamination, corrosion etc. serve as a monitor.

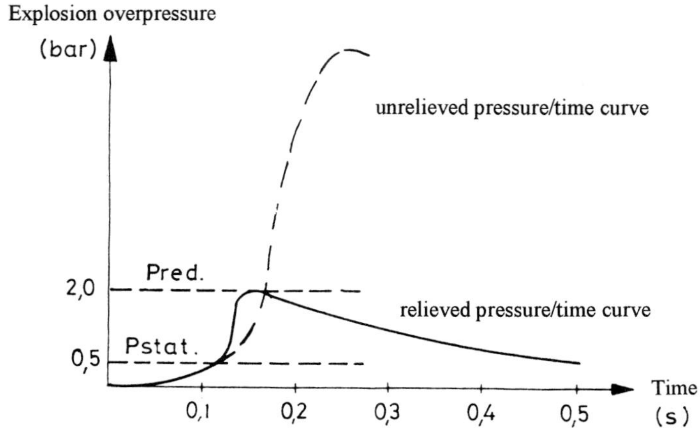

Fig. 10.16 Variation in time of pressure for a pressure relieved dust explosion
Pstat = nominal opening pressure, Pred = relieved pressure

Explosion suppression
A detector system, usually a pressure measuring device, signals an ongoing explosion and opens the valve on a pressurised vessel containing an extinguishant. Powdered or liquid extinguishant is blown into the mixer or hopper and dispersed within them. Halogenised hydrocarbons, water, powders such as ammonium phosphate and hybrid quenchers are used as extinguishants. Explosion suppression is generally effective within the range of $\mathbf{P_{max}/P_{red}}$, which also requires the equipment to be pressure shock resistant. The detectors' threshold pressure is usually around 0.1 bar. P_{red} then rises to about 1 bar. The components are accordingly designed for a pressure of 1.5 bar. The use of explosion suppression is preferred to pressure relief in the case of environmentally damaging and poisonous substances. Faultless monitoring of the detector system is required.

Protective measures in the mixing system - protection against explosion propagation
A mixing system consists of a combination of hoppers, dischargers and feeders, connecting pipework, perhaps conveyor lines, mixers, dust extraction and ventilation units, armatures and instruments. Whilst flame barrier are used in lines carrying gas, an **extinguishant barrier** serves to cut off explosion propagation in pipes carrying dust. It consists of an optical fire indicator with an automatic extinguishant sprayer. Lastly, sensor controlled fast shut off devices (gate valve) with a shut off time of less than 50 milliseconds are used in the pipework connections. Explosion protection valves can also directly react to the pressure of an explosion and be closed. Equipment such as shock resistant, explosion and flame breakthrough proof rotary valve or slide gates can be used to isolate various sections of the mixer system. Additional extinguishers are recommended to put out smouldering in product resides.

10.7.3 Combined protective measures

For reasons of technical feasibility, *the greatest possible safety* and installation and maintenance costs, several types of protective measures are often used in combination. Table 10.4 gives an overview of the combination of various protective measures (without laying any claims to being an exhaustive list):

Table 10.3 Combined protective measures for a mixing process

Open feed hopper Explosion suppression using infra red fire detectors, a safety curtain for loading the product
Feeding from containers Inertising, explosion suppression
Control sieves, crushers, delumpers Circumferential velocity <1 m/s, inertising, shock resistance
Conveying (pneumatically or mechanically) Inertising, shock resistance, explosion suppression, pneumatic conveying using an inert gas as the transporting medium
Intermediate hoppers Pressure relief, pressure proof or shock resistance with or without pressure relief, inertising, explosion suppression
Mixers Pressure proof or shock resistance with or without pressure relief, explosion suppression, inertising, circumferential velocity <1 m/s
Aspiration and ventilation equipment, filters Inertising, shock resistance with pressure relief, explosion suppression, isolation using fast shut off devices or other equipment to provide a fire barrier.

Control sieves, safety crushers, metal separators, additional grounding of all equipment, the use of antistatic materials in plastic components, filters and flexible connections serve as indirect means of preventing ignitions (sparks created by foreign bodies).

11 Implementing mixing processes

11.1 Procedure

The manifold aspects of solids mixing have been dealt with in the preceding chapters. So as to grasp the whole process with all the framework conditions encompassing it, and as is common in building any system, you need a systematic plan to translate it into an actual installation. This will be illustrated in the following and explained using actual mixing processes. The general procedure can be divided into 8 stages:

1. Formulating the tasks to be set (cf. Table 10.1) including a list of the recipes
2. Sketching out a flow diagram (see the block diagram in Fig. 11.2) to record the necessary processing steps and functions

Fig. 11.1 Batch mixing plant for production of premix, size of the mixer 3000 l, mixing time 3 minutes, product is filled into boxes and containers for further processing (extrusion).

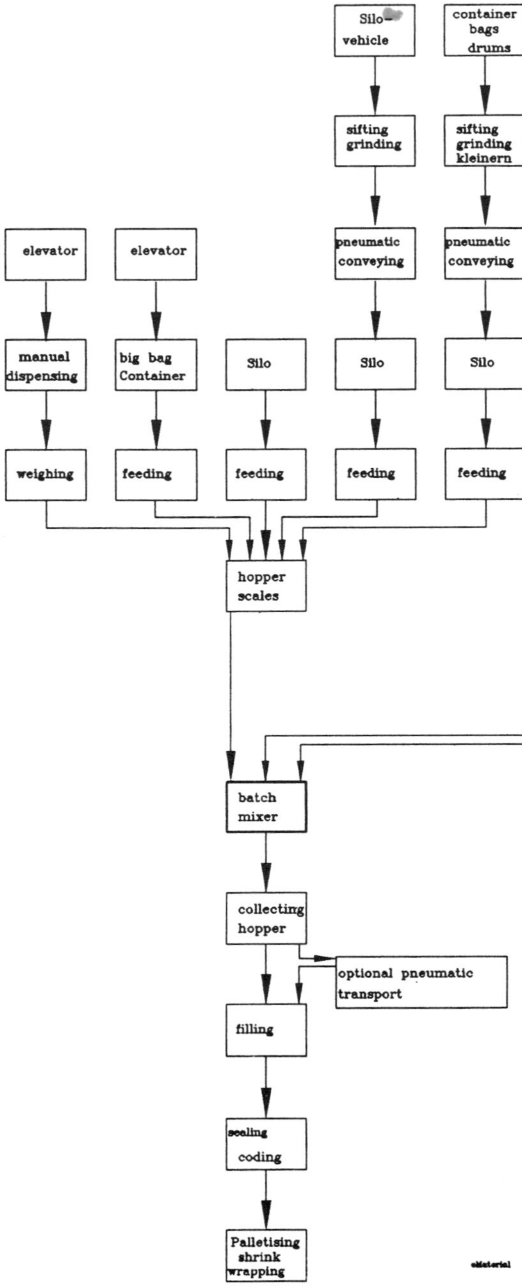

Fig. 11.2 Flowchart showing an example of a batch mixing process. Additional operations not shown: removal of any metallic particles after the ingredients have been poured off or before filling, air extraction, dust extraction and grinding

3. Drawing up a material flowchart (Fig. 11.3a, b)
   This includes the type and quantities of the materials before mixing, their packaging, packing drum, as well as the finished mixtures and filled drum. The result is a log of the material and the process (Table 11.1)
4. Deciding on the mixing process and the additional operations to be carried out by the mixer
5. Clarifying, perhaps using several variations, the nature and location of the storage, your internal transportation, your own or outside production and storage, targets to be set with regard to the delivery time for different recipes and from that the scale of the stores for ingredients and finished products, the total order delivery time, etc.
6. Tidying up the flow diagram: the building and machinery layout (the foundations)
7. Considering the system from the point of safety, with particular regard to protection against explosion and fire (cf. Chap. 10.7)
8. Estimating the investment costs. The realisable return on investment (based on the achievable production volume and sales margin) allows to fix the justified investment amounts. Besides calculating the customary pay-back period a long amortisation period of 10-15 years for the actual mixer is justified where it has been designed for universal usage.

Ingredients	1	2	3	4	5-8
Origin	in-house product	outside buy	Outside buy	in-house premix	in-house stores
Packaging	2 m^3 container	loose	200 litre barrels	1 m^3 big bag	10 litre boxes
Qty per 8 hrs	m_1	m_2	m_3	m_4	m_5-m_8
Arrival	internal, at building X	silo truck, x m^3	Lorry, loading bay, building X	internal	Pallet
Storage	container at building X	silo, y m^3	Building Y, x barrels	big bags at building X	Building Y, z m^2
Transportation	Pneumatic discharge in building y,	pneumatic	Forklift, elevator	forklift, elevator	Forklift, elevator
Additional operation	Pneumatic control sifting	Disagglomeration	Sifting		
Daily storage	5 m^3	5 m^3	2 barrels	1 big bag	X boxes
Feeding **Weighing**	Automatic	Automatic	Automatic	Automatic	fine ingredients
Mixer	batch size	Number of mixers	and	Batches	

Fig. 11.3a Example of a study of the flow of material up to a batch mixer

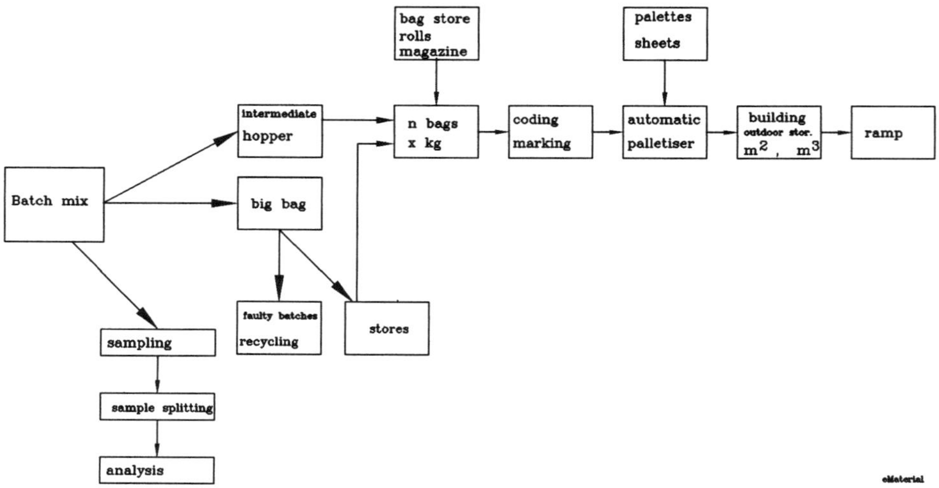

Fig. 11.3b Example of a material flowchart for the finished mix produced in a batch mixer

11.2 The Batch Mixing Process

A hierarchy of operational safeguards must be observed for reasons of quality and product liability. This gives rise to an order of precedence with regard to guaranteeing the system's quality, technical design and availability:

1. *The mixer*: The primary criterion is the final mixture's homogeneity. When purchasing the mixer, compromising on its technology at the risk of producing non-uniform mixes has irreparable consequences. Even perfect feeding, storage and process automation cannot correct these faults.
2. *Avoiding segregation up to the time of final use of the mixture*
3. *Feeding of the ingredients plus weighing, creating a report and storage of the data*
4. *The mixing system's throughput*: too low a throughput volume can be balanced out by longer production periods, e.g. in the case of recipes which take time to prepare.
5. *Storage of the ingredients, availability for delivery*

Table 11.1 Material and process record

Input ingredients	
Origin	
Packaging	
Required quantity/ unit of production	
Arrival: location, coding, sample	
Stores: reception store, buffer stores	
Transportation to the mixer system	
Additional operations: sieving, removal of foreign bodies, disagglomeration, (coarse) crushing	
Day's storage in the mixer system	
(Delivery) feeding, weighing	
Mixing	
Logistics, data management, reporting, coding, process control, (CIM/CIP pyramid)	
Products (mixtures)	
Quantity per hour, per 8 hours etc.	
Mixer (batch size, number)	
Sampling, inspection	
Intermediate store (hoppers, containers, etc.)	
Acceptance and storage of faulty batches	
Filled packing drum: type, store, delivery of packing drum	
Filling, sealing, packing, palleting, coding	
Store	
Delivery	
Dust extraction:	ingredients, filters and filter dust, where, quantity/8 hours, filter dust usage
Cleaning:	machine/ hopper/building cleaning frequency type of cleaning
Quality control:	sampling and analysis of the finished ingredients and mix removal of foreign bodies control sifting taking samples from the filled drums faulty batch procedure

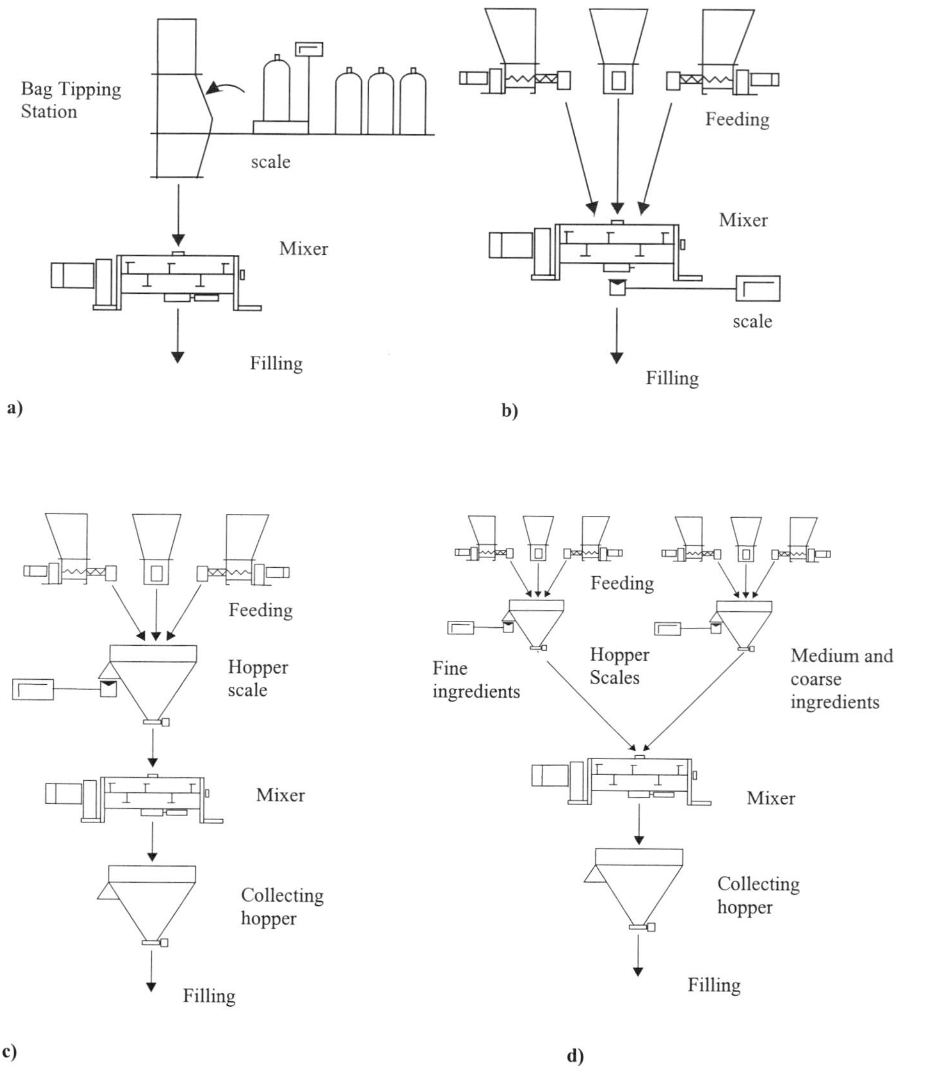

Fig. 11.4 Influence of the system's configuration on throughput (in ascending order from a to d)
a) Ingredients are weighed by hand and tipped into the mixer; **b)** ingredients are fed automatically into the mixer which serves as a set of hopper scales; **c)** hopper scales are placed above the mixer and a collecting hopper below it. A higher number of mix cycles can be achieved by the reduction in the mixer's filling and emptying times; **d)** two hopper scales (for coarse and medium ingredients and for fine substances) reduce the feeding and weighing time when preparing a recipe and at the same time increase the accuracy of the weighing.

Fig.: 11.5 PVC- Mixing: Installation consisting of two 1200 l batch mixers (background). The mixers are filled by sack dipping and pneumatic transport or directly from Silos with pneumatic transport (not shown). Liquid ingredients are added via 2-phase-nozzles. Mixers discharge onto transport belts which supply the mixture to the calendars.

Each degree of automation must observe the regulations governing **safety, workers' protection and environmental protection**. In the choice of the mixing process and mixer, the legal provisions applied to filling packing drums for commercial sale are not dealt with here, neither is the question of extracting the air out of the material before or during filling to save on drum storage and transportation costs.
On the basis of the above order of priorities, cost savings on the mixer itself is bottom of the list. It would be preferable to add on the peripheral devices for feeding, weighing, storing etc. in stages. Criteria for choosing the mixer itself are to be found in Chap. 10.

The influence of the installation's configuration on throughput is shown in the flow charts in Fig. 11.4 (a-d). The systems present an increasing throughput from a to d (in each case using the same design of mixer). The design of the mixer, whether horizontal or vertical, and the layout as shown in Fig. 11.4 fundamentally influence the height and surface area requirement. The space requirements of a mixing system can be significantly altered by *lifting* the ingredients up to the mixer or by *lifting* the finished mix (without segregation) after it leaves the mixer. Likewise, the process determines the system's configuration as is shown for a mix/grinding/mix process in Fig. 11.6. The type and volume of storage for the ingredients in silos, big bags, sacks or containers are criteria governing the required space.

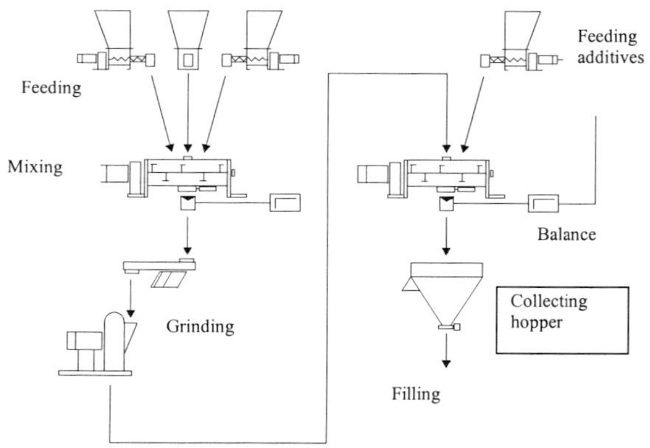

Fig. 11.6 Mix/grinding/mix process using two mixers in series. Additives are being fed into the second mixer.

11.3 Typical Configurations of Batch Mixing Systems

Figs. 11.7 to 11.10 show configurations of batch mixing systems for various tasks.

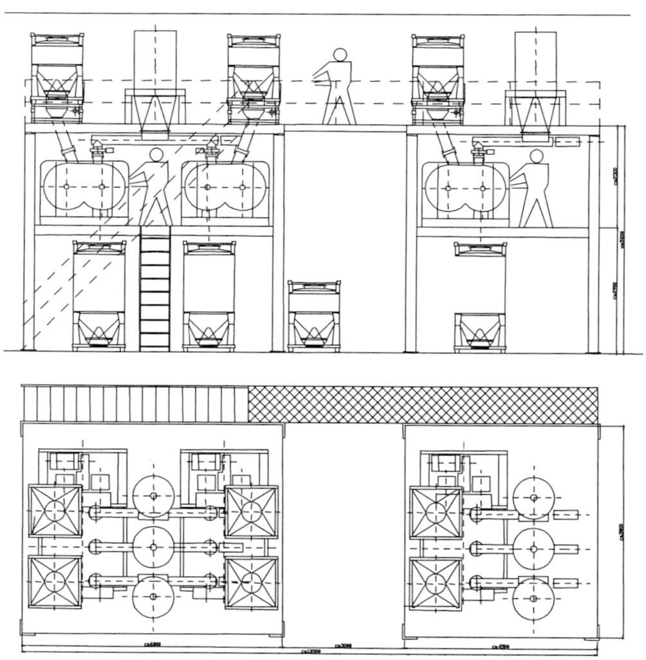

Fig. 11.7: Batch mixing installation with container supply of the ingredients and container receiving hoppers. Due to ease of transport as well as cleaning the containers are used especially in food and

pharmaceutical productions. In this example the 3 principal components are stored in intermediate hoppers which discharge by screw feeders into the 2 mixers.

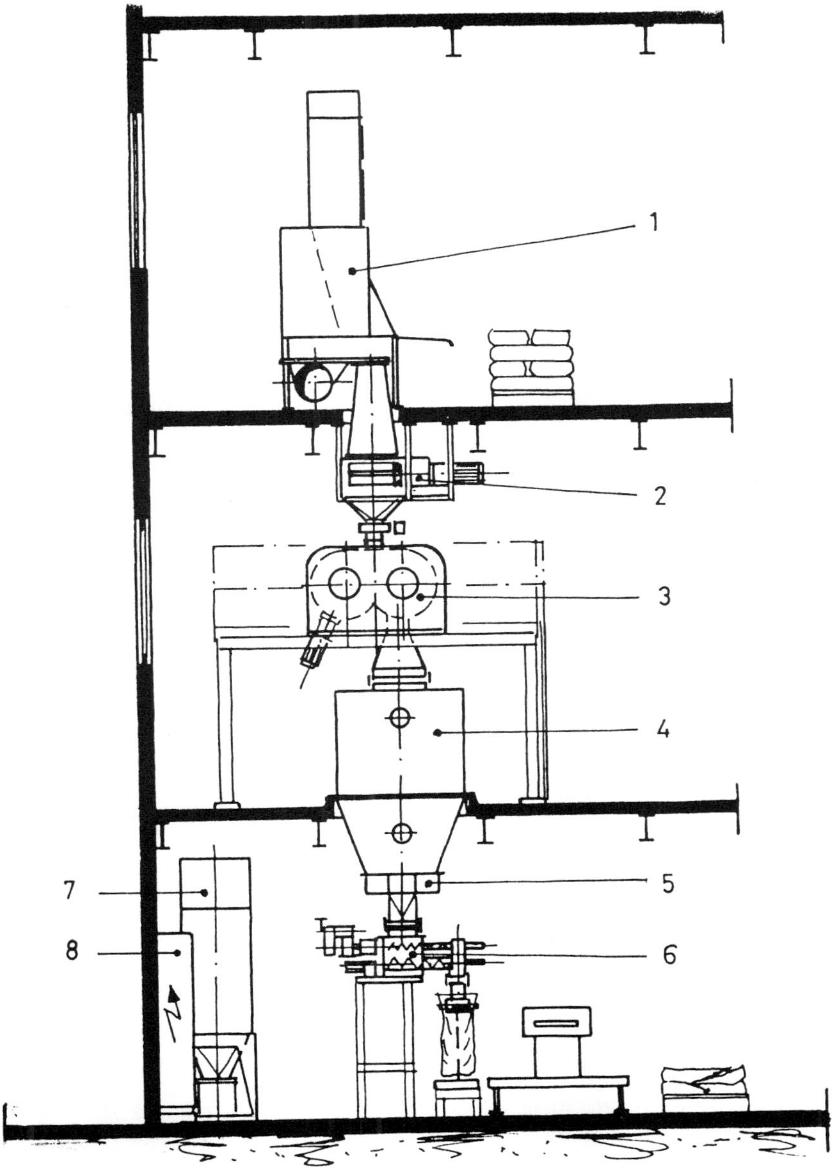

Fig. 11.8 Vertical mixing installation: 1 Bag dipping and compactor; 2 Coarse crusher/nibbler; 3 Mixer; 4 Collecting hopper; 5 Bin discharger; 6 Filling station; 7 main dust extractor; 8 Control pannel

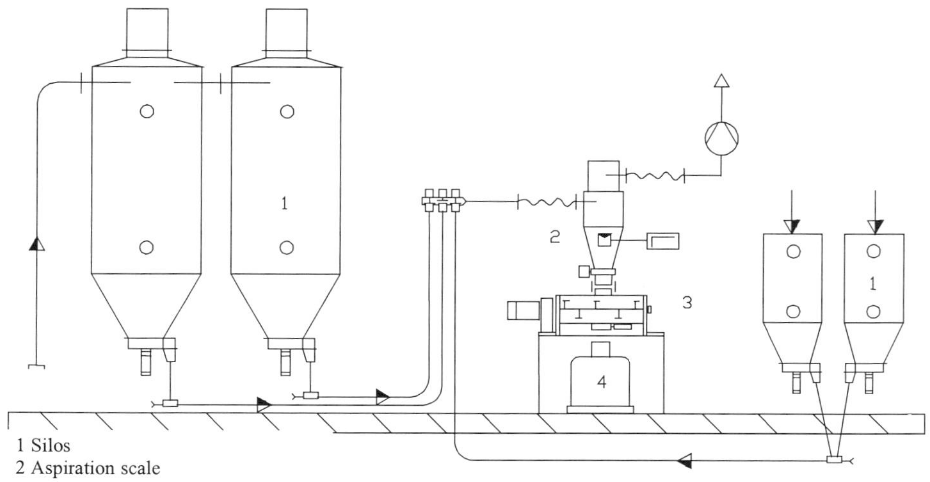

1 Silos
2 Aspiration scale
3 Mixer
4 Filling

Fig. 11.9 Feed of the components to the mixer by pneumatic conveying

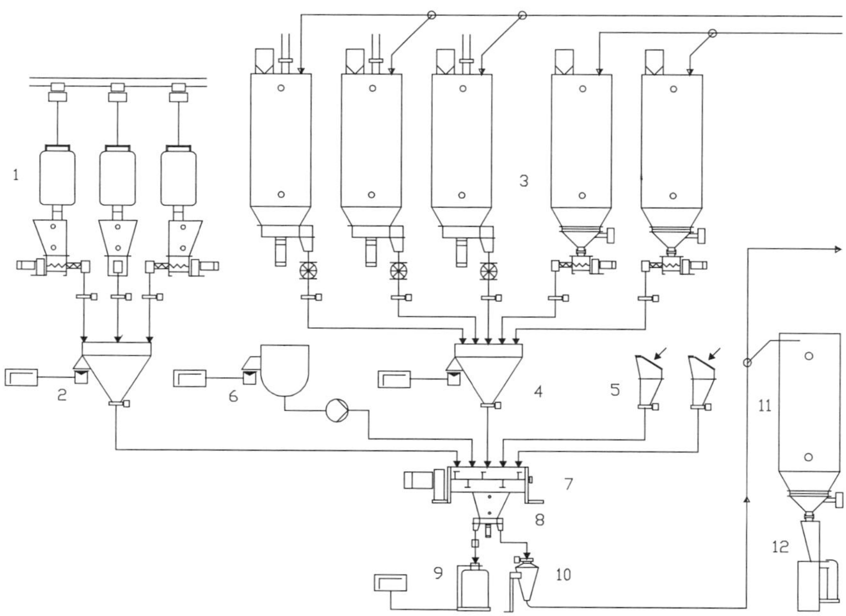

Fig. 11.10 Batch mixing system with coarse and fine ingredient hopper scales and direct feed. The product is lifted pneumatically using dense phase conveying (without the mix segregation). 1 Fine ingredients in big bags with a feeder; 2 Hopper scales for fine ingredients; 3 Coarse and medium ingredients with silos, discharge device and feeder; 4 Hopper scales for coarse and medium ingredients; 5 Manual feed for special substances; 6 Addition of liquids from hopper scales; 7 Mixer; 8 Collecting hopper; 9 Direct filling; 10 Pneumatic dense phase conveying of the mix produced; 11 Intermediate store; 12 Filling line

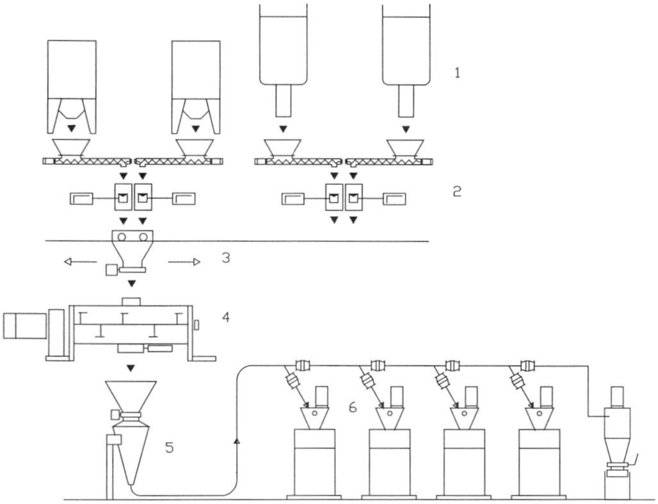

Fig. 11.11 Coarse and fine ingredients are weighed individually because of the exacting demands made on the feeding system.
1 Ingredient drum with feeder; 2 Individual hopper scales; 3 Travelling collecting hopper for picking up the weighed ingredients; 4 Mixer; 5 Pneumatic dense phase conveyor; 6 Filling or tabletting machine

11.4 The Continuous Mixing Process

The procedure is similar to that for a batch mixer (Chap. 11.1 & 2). You have to decide whether, through premixing them batchwise, it is possible to reduce the number of ingredients that form only a small percentage by volume. In selecting the feeding system you have to specify the feeding's accuracy, detailing the time interval and output range for which this accuracy is to be maintained. Here it is important whether the feeding accuracy is to be observed for the lowest or average feed rate of each ingredient, especially in the case of minor ones (cf. Chap. 10). The throughput can be configured for an average bulk density of the end materials or a maximum volumetric throughput at their lowest bulk density. A quality inspection by off or in-line analyses should be provided for at the mixer's outflow.

Fig. 11.12: Continuous mixing of PVC and other components. 5 Loss-in-weight feeders are feeding at constant rate onto a belt which supplies the material to the compact continuous mixer with a capacity of 1200 l/h. The feeders are refilled by big bags (not shown).

Extreme accuracy where the ingredients have low feed rates or extremely high throughputs for materials with possibly low bulk densities means more expense, which perhaps is not justified in the average production case.

11.5 Typical configurations of continuous mixing systems

Figs. 11.15 to 11.16 show examples of configurations of continuous mixing systems:

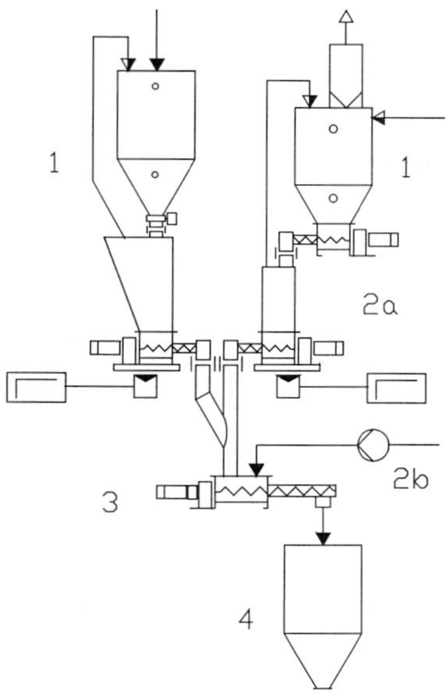

Fig. 11.13 Continuous feeding and solids/liquid mixing system. 1 Ingredient silos; 2a Loss-in-weight feeders for solids; 2b Feed pump eventually by loss-in-weight feeders for liquids; 3 Continuous mixer; 4 Collecting hopper and filling.

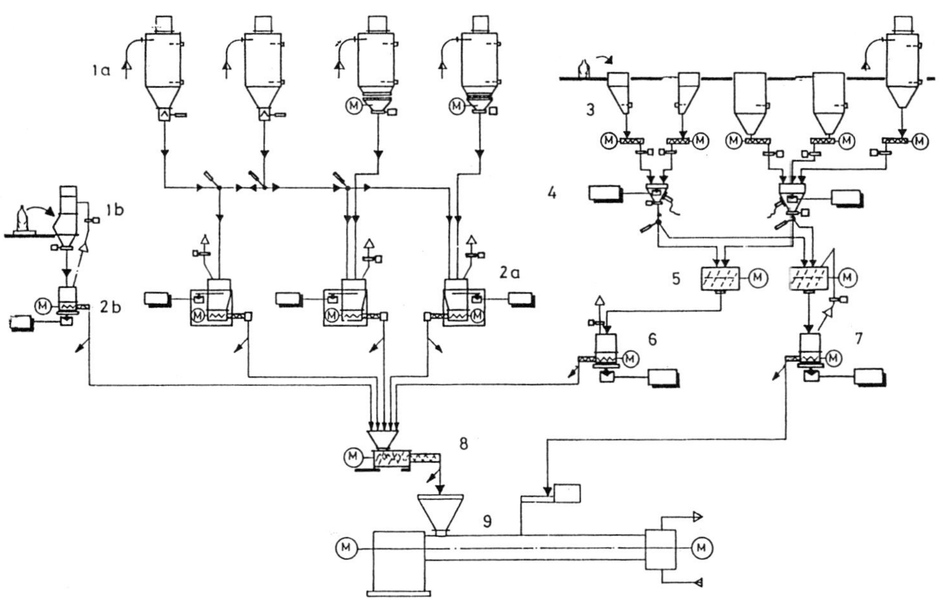

Fig. 11.14 Continuous feeding, mixing and **extrusion** for plastic compounding using prepared premixes in batch mixers. 1a Ingredient silos and feeder units; 1b Ingredient sack dipping; 2a,b Loss-in-weight feeders; 3 Batch feeding; 4 Hopper scales for premix; 5 Premix batch mixer; 6 Loss-in-weight feeders for the continuous addition of the premix to the main ingredients; 7 Los-in-weight feeders for adding the premix to the extrusion melt (split feed); 8 Continuous mixer; 9 Extruder

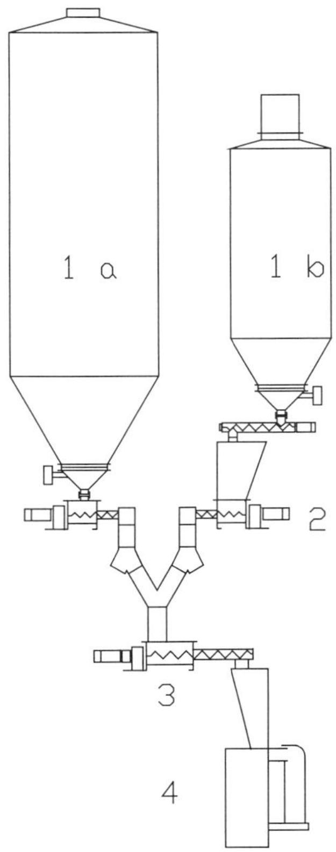

Fig. 11.15 It is only possible to achieve adequate mixing of ingredients with extremely differing bulk densities (0.02 kg/l and 1.0 kg/l) in a continuous mixing process. 1a,b Ingredient silos; 2 Volumetric feeding; 3 Continuous mixer; 4 Filling

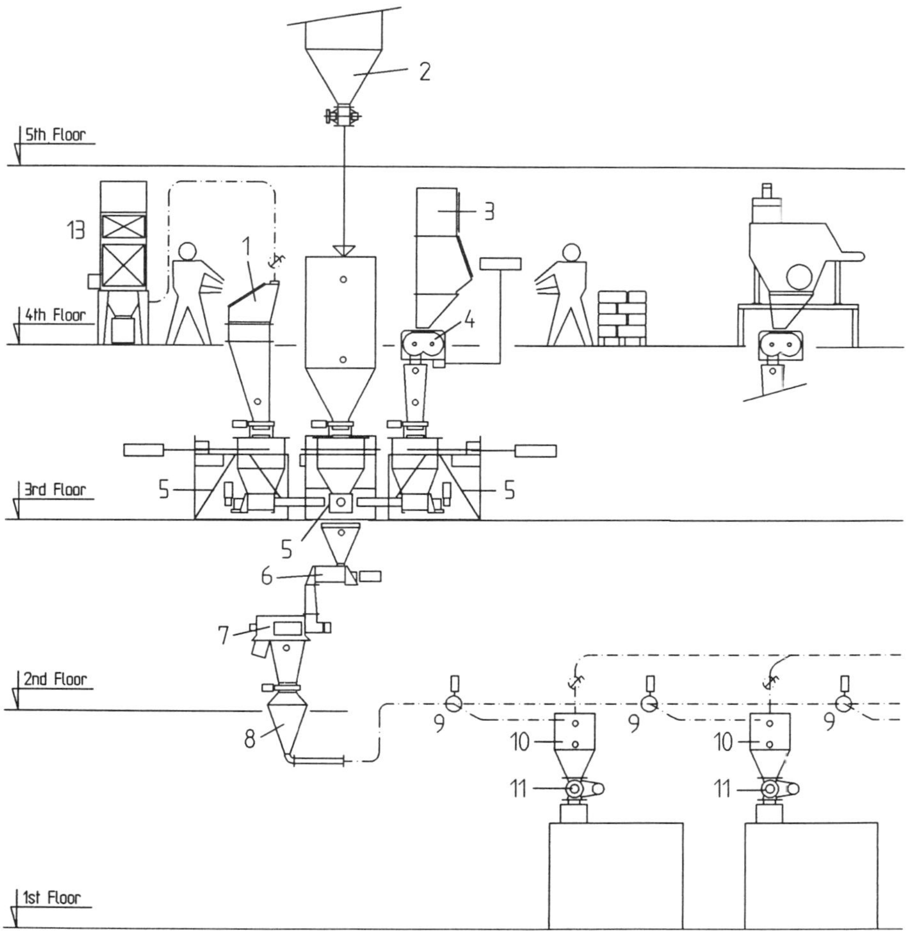

Fig. 11.16 Continuous feeding and mixing system for food products with pre-mix batch mixing
1 Bag emptying station with vibrating control sieve; 2 Main ingredient silo; 3 Bag emptying filter for minor ingredients; 4 Gericke multi-flux batch mixer for pre-mix; 5 Loss-in-weight feeders with system control; 6 Continuous mixer; 7 Centrifugal sifter; 8 Dense phase conveyor; 9 Diverter valves; 10 Receiving hoppers above packing machines; 11 Rotary valves; 12 Dust collection.

Bibliography

[1] *R. Weinekötter, H. Gericke:* Mischen von Feststoffen, Springer-Verlag, Berlin, 1995

[2] *V.W. Uhl, J.B. Gray*: Mixing, Vols. 1-3, Academic Press, Orlando, USA, 1986

[3] *J.C. Williams*: Mixing, Theory and Practice, Vol. 3, edited by V.W. Uhl and J.B. Gray, Academic Press, Orlando, 1986

[4] *P.V. Dankwerts*: The Definition and Measurement of Some Characteristics of Mixtures, Appl. Sci. Research, 1952, Ch.. 3, p. 279ff

[5] *A. Babl, W. Steiner*: "Siliciumcarbid" [Silicon Carbide], the Chap. in Ulmanns Encyclopädie der Technischen Chemie [Ulmann's Encyclopaedia of Industrial Chemistry] Vol. 21, Verlag Chemie, Weinheim, 1982

[6] *J. Raasch*: Notes for the course of lectures on "Mischen und Rühren" [Mixing and Stirring], summer semester 86, University of Karlsruhe (Technical College)

[7] *J. Raasch, K. Sommer*: Anwendung von statistischen Prüfverfahren im Bereich der Mischtechnik [The application of statistical test procedures in the field of mixing technology] Chemical Engineering 62 (1990) No. 1, pp. 17-22

[8] *E. Kreyszig*: Statistische Methoden und ihre Anwendungen [Statistical methods and their applications], 7th edition, Vandenhoeck und Ruprecht, Göttingen, 1985

[9] *R. Weinekötter, L. Reh*: Continuous Mixing of Fine Particles, Part.Part. Syst. Charact. 12 (1995) pp. 46-53

[10] *R. Holzmüller*: Untersuchung zur Schüttgutbewegung beim kontinuierlichen Feststoffmischen [Investigation into the movement of bulk material during continuous solids mixing], dissertation, University of Stuttgart, 1984

[11] *H. Egermann, N.A. Orr*: Comments on the paper "Recent developments in solids mixing" by L.T. Fan et alia; Powder Technology 68 (1991) pp. 195-196

[12] *L.T. Fan, Y. Chen, F.S. Lai*: Recent developments in solids mixing; Powder Technology 61 (1990) pp. 255-287

[13] *P.M.C. Lacey*: The Mixing of Solid Particles, Trans. Instn. Chem. Engrs. 21 (1943) pp. 53-59

[14] *K. Stange*: Die Mischgüte einer Zufallmischung als Grundlage zur Beurteilung von Mischversuchen [The mix quality of a random mix as the basis for evaluating mixing trials], Chemical Engineering 26 (1954) No. 6, pp. 331-337

[15] *K. Sommer*: Probenahme von Pulvern und körnigen Massengütern [Taking samples of powder and granular bulk goods], Springer Verlag, 1979

[16] *K. Stange*: Angewandte Statistik [Applied Statistics], Springer Verlag, Berlin, Heidelberg and New York, 1971

[17] *J.S. Bendat, A.G. Piersol*: Random Data - Analysis and Measurement Procedures, 2nd edition, John Wiley & Sons, New York, 1986

[18] *J.R. Bourne*: Variance sample size relationships for incomplete mixtures, Chemical Engineering Science, 1967, Vol. 22, pp. 693-700

[19] *K. Stange*: Genauigkeit der Probenahme bei Mischungen körniger Stoffe [Granular mix sampling accuracy], Chemical Engineering 30, 1967 Vol. 9/10, pp. 582-592

[20] *P.M. Lacey, F.S. Mirza*: A Study of the Structure of Imperfect Mixtures, Parts I & II, Powder Technology 14 (1976) 17/24, 25/33

[21] *H.P. Geering*: Meß- und Regelungstechnik [Measurement and Control Technology], Springer Verlag, Berlin, 1988

[22] *K. Sommer*: Wie vergleicht man die Mischfähigkeit von Feststoffmischern [How to compare the mixing properties of solids mixers], Preparation Technology No. 5 (1982), pp. 266-269

[23] *W. Entrop*: International Symposium on Mixing, D1, 1-14, Mons, 1978

[24] *W. Müller*: Untersuchungen über Mischzeit, Mischgüte und Arbeitsbedarf in Mischtrommeln mit rotierenden Mischelementen [Studies on the mixing time, mix quality and effort in respect of mixer drums with rotating mixer tools], dissertation, Karlsruhe Technical College, 1966

[25] *W. Müller, H. Rumpf*: Das Mischen von Pulvern in Mischern mit axialer Mischbewegung [Mixing powders in axial motion mixers], Chemical Engineering 39, (1967) pp. 365-373, Weinheim

[26] *W. Müller*: Methoden und derzeitiger Kenntnisstand für Auslegungen beim Mischen von Feststoffen [Methods and the current state of the art in solids mixing configurations], Chemical Engineering 53 (1981), pp. 831-844

[27] *K. Sommer*: Mechanismen des Pulvermischens [The mechanics of powder mixing], Chemical Engineering 49 (1977), pp. 305-311

[28] *R. Miyanami*: Chapter 2 of Mixing and Kneading Technology, (edited by T. Yano), Nikan Kogyo Shinbun Sha, 1980, text in Japanese

[29] *Ph. B. Rudolf von Rohr, F. Widmer*: Scaling up and homogeneity investigations of ribbon mixers, lecture to the 1st World Congress on Particle Technology, Nürnberg, April 16-18, 1986

[30] *R. Weinekötter*: Kontinuierliches Mischen feiner Feststoffe [The continuous mixing of fine solids], dissertation No. 10083, Zurich ETH, 1993

[31] *W. Müller*: Energiebedarf beim Homogenisieren im pneumatischen Mischbunker [Energy requirement of the homegenising process in a pneumatic mixing silo], Chemical Engineering 46, 1974, No. 7, p. 295

[32] *J. Werther*: Strömungsmechanische Grundlagen der Wirbelschichttechnik [Principles of fluidisation flow mechanics], Chemical Engineering 49 (1977), No. 3, pp. 193-202

[33] *J. Schmalfelld*: VDI-Z 118 (1976) No. 2, p. 65/72

[34] *L. Reh*: Wirbelschichtreaktoren für nichtkatalytische Reaktionen [Fluidising reactors for non-catalytic reactions], Chapter in Ulmanns Encyclopädie der technischen Chemie [Ulmann's Encyclopaedia of Industrial Chemistry], 4th edition, Vol. 3, 1973

[35] *J. Schwedes, J. Otterbach*: Dimensionierung von pneumatischen Granulatmischern [Dimensioning pneumatic granulating mixers], Process Technology 8 (1974), No. 2

[36] *H. Gericke*: Dosieren und Mischen von Schüttgütern im Chargen- und kontinuierlichen Betrieb [Feeding and mixing bulk goods in batch and continuous operation], Aufbereitungstechnik 22 (1981), Bk. 1, pp. 15-21

[37] *J.R.. Bourne*: Chapter 8 of the course of lectures on "Chemische Reaktionstechnik I/II [Chemical reactions technology I/II], Zurich ETH, 1985

[38] *C.Y. Wen, L.T. Fan*: Models for Flow Systems and Chemical Reactions, Marcel Dekker Inc., New York, 1975

[39] *W. Bartknecht*: Explosionsschutz [Anti-explosion protection], Springer Verlag, Berlin, 1993

[40] *L.T. Fan et alia*: Evaluation of a Motionless Mixer Using a Radioactive Tracer Technique, Powder Technology 4 (1970/71), pp. 345-350

[41] *O. Molerus*: Über die Axialvermischung bei Transportprozessen in kontinuierlich betriebenen Apparaturen [Axial mixing in the transportation processes in continuously operated plant] Chemical Engineering 38 (1966), Vol. 2

[42] *O. Levenspiel*: Chemical Reaction Engineering, 2nd edition, John Wiley & Sons, New Yorrk, 1972

[43] *M. Weber*: Dispersionsmodell für Mischer [Dispersion model for mixers] diploma work at the Zurich ETH's Institute for Processing and Refridgeration Engineering, December 1991

[44] *P.V. Dankwerts*: Continuous Flow Systems. Distribution of Residence Times, Chemical Engineering Sciences 1953, No. 2, p. 1

[45] *M. Tesch, L. Reh*: Zeitkonstantes Dosieren feiner Feststoffe [Time constant feeding of fine solids], Chemical Engineering 64 (1992), No. 11, pp. 1034-1036

[46] *J.C. Williams R. Richardson*: The Continuous Mixing of Segregating Particles, Powder Technology 33 (1982), pp. 5-16

[47] *l. Tschuor*: Charakterisierung eines Doppelwellenrührwerkes [Characterisation of a dual shaft agitator], dissertation 9425, Zurich ETH, 1991

[48] *R.H. Wang*: Residence Time Distribution Models For Continuous Solids Mixers: proceedings - Powder and Bulk Solids Handling and Processing, 11th Annual Powder and Bulk Solids

Conference and Exhibition, pp. 8-17, Rosemont, Illinois, USA, 1986

[49] *K. Borho et alii:*Product Properties and Processing, Video, BASF, Ludwigshafen, BASF DCE 60

[50] *K. Sommer*: Mixing of Solids, in Ulmann's Encyclopaedia of Industrial Chemistry, Vol. B4, Chapter 27, VCH Publishers Inc., 1992

[51] *P. Müller*: Partikel-Widerstand und Druckabfall in beschleunigter Gas-Feststoff-Strömung [Particle resistance and pressure drop off in an accelerated gas/solids flow], dissertation, Zurich ETH, 1992, 9842

[52] *J.C. Williams*: Fuel Soc. Journal, University of Sheffield 14, 29, 1963

[53] *B. Scarlett*: Particle Process Control, plenary PARTEC session, Nürnberg, 24-26th March 1992

[54] *Phil Williams, Karl Norris (editor)*: Near-infrared technology in the agricultural and food industries, St. Paul, Minnesota: American Association of Cereal Chemists, 1987

[55] *Terence Allen*: Particle size Measurement, 4th edition, Chapman and Hall, London, 1990

[56] *B. Stalder*: Ermittlung der Mischgüte beim Pulvermischen [Establishing mix quality when blending powders], dissertation 10236, Zurich ETH, 1993

[57] *S.S. Weidenbaum*: "Mixing of Solids" in Advances in Chemical Engineering, Vol. 2, edited by T.B. Drew and J.W. Hoopes, Academic Press Inc., New York, 1958

[58] *B.H. Kaye, A. Brushenko, R.L. Ohlaber, D.A. Pontarelli*: A fibre optics probe for investigating the internal structure of powder mixtures, Powder Technology 2 (1968/69), pp. 243-245

[59] *C.F. Harwood, R. Davies, M. Jackson, E. Freeman*: An Optic Probe for Measuring Mixture Composition of Powders, Powder Technology 5 (1971/72), pp.77-80

[60] *M. Alonso, M. Satoh, K. Miyanami*: Recent works on Powder Mixing and Powder Coating Using an Optical Measuring Method, KONA No. 7 (1989), pp. 97-104

[61] *R. Weinekötter, R. Davies, J.C. Steichen*: Determination of the Degree of Mixing and the Degree of Dispersion in Concentrated Suspensions: proceedings of the Second World Congress, Particle Technology, pp. 239-247, 19th-22nd September 1990, Kyoto, Japan

[62] *R. Weinekötter, L. Reh*: Characterization of Particulate Mixtures by In-line Measurements, Part. Syst. Charact. 4/94

[63] *L. Reh, J. Li*: Measurement of Voidage in Fluidized Beds by Optical Probes, proceedings at the 3rd International Conference on Circulation Fluidized Beds, Nagoya, Japan, Oct. 1990, pp. 4-16-1 -9

[64] *K. Stange*: Angewandte Statistik {Applied statistics] 1, Springer Verlag, Berlin, 1970

[65] *J. Raasch, K. Sommer*: Anwendung von statischen Prüfverfahren im Bereich der Mischtechnik [Application of statistical test procedures in the field of mixing], Chemical Engineering 62 (1990) No. 1, pp. 17-22

[66] *H.B. Ries*: Mischtechnik und Mischgeräte [Mixing and mixers], Preparation Technology, Bks. 1 + 2, 1979

[67] *H-P. Wilke, R. Buhse, K. Groß*: Mischer - Verfahrenstechnische Grundlagen und apparative Anwendungen [Mixers - process technology principles and equipment applications], Vulkan Verlag, Essen, 1991

[68] *K. Sommer, G. Hauser, H. Alexy*: Gutachten über den Gericke Mehrstromfluidmischer [Expert opinions on the Gericke multi-stream fluids blender], Weihenstephan, 1986

[69] *G. Scheuber*: Untersuchung des Mischungsverlauf in Feststoffmischern unterschiedlicher Größe [Investigation into the mixing process in solids mixers of varying sizes], dissertation, University of Stuttgart, 1979

[70] *G. Dau, F. Ebert, F. Hähner*: On-line Analyse der Austrittskonzentration von Feststoffmischern [On-line analysis of the outflow concentration of solids mixers], Preparation Technology 35, 1994, No. 6, pp. 281-289

[71] *T. Koch, K. Sommer*: Modelling of Continuous Granulation in a Fluidized Bed, proceedings at 1st Particle Technology Forum, Part 1, pp. 17-184, Denver, 1994

[72] *R. Meili*: Scale-up von Feststoffmischern [Scaling up solids mixers], semester work at the Zurich ETH's Institute for Process Technology and Refridgeration Engineering, WS 88/89

[73] *J. Villermaux*: Génie de la réaction chimique [The nature of chemical reactions], Lavoisier, Paris, 1982

[74] *K. Sommer*: Continuous Powder Mixing, proceeding at the 1st Particle Technology Forum, Part III, pp. 343-349, Denver, 1994

[75] *A.H. Lefebvre*: Atomization and Sprays, Taylor & Francis, Bristol PA, USA, 1989

[76] *T. Isenschmid, P. Schmid, L. Reh*: Continuous Fluid Bed Granulation and Atomization with Industrial Two-Fluid Nozzles, Preprints 1, European Symposium on Particle Measurement Technology (PARTEC), Nürnberg, Germany, March 1992

[77] *T. Gamma*: Vergleich einer innen-mit einer aussenmischenden Zweistoffdüse [Comparison of an internal with an externally mixing twin ingredient jet], semester work in "Mechanical process technology" at the Zurich ETH's Institute for Process Technology and Refridgeration Engineering, February 1994

[78] *F. Briegmann*: from Ulmann's Encyclopaedia of Industrial Chemistry, 4th edition, Vol. B2, p. 321ff, Verlag Chemie, Weinheim, 1974

[79] *K. Sommer*: Ulmann's Encyclopaedia of Industrial Chemistry, 5th edition, Vol. B2, Chap. 7.2, VCH Verlagsgesellschaft, Weinheim, 1988

[80] Wärmeübergang von einer Heizfläche an ruhende oder mechanisch durchmischte Schüttungen [Heat transference from a hot surface to stationary or mechanically mixed bulk], excerpt from the VDI Heat Atlas, 8th edition, Springer Verlag, Berlin, 1997

[81] *H. Gericke*: Dosieren von Feststoffen (Schüttgütern) [Solids feeding (bulk materials)], Gericke GmbH (editors), Rielasingen, 1989

[82] *G. Vetter*: Die Dosiergenauigkeit bei der Stoffdosierung [Feed accuracy in material feeding], Chemical Engineering 61 (1989), No. 2, pp. 136-140

[83] *K. Hlavica*: Differentialdosierwaagen [Differential feeding scales] in the Handbook of Feeding, edited by G. Vetter, Vulkan Verlag Essen, 1994

[84] *E. von Haefen and H-G. Schecker*: DUSTEXPERT - Ein Expertensystem zur Beurteilung von Explosionsgefahren und zur Auswahl von Explosionsschutzmassnahmen bei staubverarbeitenden Anlagen [An expert system for evaluating the risks of explosion and for selecting anti-explosion countermeasures for dust processing plant], VDI Report 975, Sichere Handhabung brennbarer Stäube [The safe handling of combustible dusts], Meeting of Nov. 1992. The VDI and DIN Commission for the prevention of air pollution, VDI Verlag, Düsseldorf, 1992, pp. 359-378

[85] VDI Report 975, Sichere Handhabung brennbarer Stäube [The safe handling of combustible dusts], Meeting of Nov. 1992, the VDI and DIN Commission for the prevention of air pollution, VDI Verlag, Düsseldorf, 1992, pp. 359-378

[86] *Bartknecht, W*: Staubexplosionen, Ablauf und Schutzmaßnahmen [Dust explosions, their course and preventative measures] Springer Verlag, Berlin, Heidelberg, 1987

[87] VDI guideline 2263: Staubbrände und Staubexplosionen. Gefahren-Beurteilung-Schtzmaßnahmen. Blatt 1 Untersuchungsmethoden zur Ermittlung von sicherheitstechnischen Kenngrößen von Staäuben. [Dust fires and explosions. Risk evaluation and preventative measures. Sheet 1 of "Investigative methods for establishing the safety values of dusts"] Beuth Verlag, Berlin, May 1990

[88] VDI guideline 2263 Sheet 2: Inertisierung [Inertising], May 1992

[89] VDI guideline 2263 Sheet 3: Explosionsdruckstossfeste Behälter und Apparate, Berechnung, Bau und Prüfung [Blast resistant containers and apparatus. Calculations, construction and testing], May 1990

[90] VDI guideline 2263 Sheet 4: Unterdrückung von Staubexplosionen [Dust explosion suppression], April 1992.

[91] Safety information and worksheets 140260-140279 from the BIA handbook on Brenn-und Explosions-Kenngrössen von Stäuben [Combustion and explosion values of dusts], Berufsgen. Institute für Arbeitstechnik (editors), Erich Schmidt Verlag, Bielefeld, 1987

[92] *F. Schmalz*: Sicherheit chemischer und verfahrenstechnischer Anlagen [The safety of chemical and process control systems], course of lectures at the Zurich ETH, summer semester1994

[93] *C. Donat*: Explosionsfeste Bauweise [Explosion proof construction]. Report to the International Committee for the Prevention of Accidents at Work and Occupational Illnesses in the Chemical Industry, Lucerne, 1984

[94] VDI guideline 3673. Druckentlastung von Staubexplosionen [Dust explosion pressure relief], Oct. 1983, Draft amendment Nov. 1992

[95] *W. Pietsch*: Size Enlargement by Agglomeration, John Wiley & Sons/Salle + Sauerländer, Chichester, UK/Aarau, Switzerland/Frankfurt am Main, 1992

[96] *F. Hoornaert, G. Meesters, S. Pratsinis, B. Scarlett*: Powder Agglomeration in a Lödige Granulator, proceedings at the 1st Particle Technology Forum, part I, pp. 278-286, Denver, 1994

[97] *H.G. Kristensen*: Particle Agglomeration in High Speed Mixers, proceedings at the 1st Particle Technology Forum, part I, pp. 214-219, Denver, 1994

[98] *W. Pietsch*: Parameters to be Considered during the Selection, Design and Operation of Agglomeration Systems, proceedings at the 1st Particle Technology Forum, part I, pp. 248-257, Denver, 1994

[99] *NAMUR (Normen- und Arbeitsgemeinschaft für Meß- und Regeltechnik in der chemischen Industrie* - Standards and chemical industries measurement and control technology study group*)*: Recommendation Ak. 3.3: Dosiergenauigkeit von kontinuierlichen Waagen, [The feed accuracy of continuous scales], Leverkusen, July 1993

[100] *J.Wunschmann*: Wärmeübertragung von beheizten Flächen an bewegte Schüttungen bei Normaldruck und im Vakuum, PHD-thesis University of Karlsruhe, 1974

[101] *E.Tsotsas, E.U. Schlünder*: Contact drying of mechanically agitated particulate material in the presence of an inert gas, Chem. Eng. Process. 20 (1986), p. 277/285

[102] *R. K. Eckhoff*: Dust Explosions in the Process Industries, Butterworth, 1997

[103] *J. F. Zevenbergen, A. E. Dahoe, A.A. Pekalski, B. R. de Ruiter, B. Scarlett*: Laser ignition of single magnesium particles

[104] *H.Wilms*: Performamce of gravity flow blending silos, The First European Congress on Chemical Engineering, Florence, Italy 1997, Proceedings vol. 2, p.971-974

[105] *Eichler and Dau*: Geometry and mixing of gravity discharge silo mixers, The First European Congress on Chemical Engineering, Florence, Italy 1997, Proceedings vol. 2, p.971-974

[106] *L. T. Fan*, F. S. Lai: A Study on the Mixing of Flour in a Motionless Sulzer (Koch) Mixer Using Radioactive Tracer, Powder Techn. 13 (1976)

[107] J. *Gyenis, J. Nemeth:* Two Phase (Gas-Solid) Particle Flow in Vertical Mixers Tubes Equipped with Motionless Mixer Elements, AICHE First Int. Particle Technology Forum 1994, Part III, p.454-459

[108] *O. Eichstädt*: Kontinuierliches Mischen Feiner Feststoffen in Fluid-Dynamischen Fallrohrmischern (Continuous Mixing of Fine Particles within Fluid Dynamic Vertical Tube Mixers), Dissertation ETH-Zurich 1997 by Gravity Fall

[109] *R. Weinekötter*: How to avoid overmixing, Chemical Plants and Processing, November 1994, p.48-p.50

Subject Index

PARTICLE TECHNOLOGY SERIES

FORMERLY POWDER TECHNOLOGY SERIES

1. H. Rumpf (Translated by F.A. Bull): *Particle Technology*. 1990 ISBN 0-412-35230-3
2. S. Lowell and Joan E. Shields: *Powder Surface Area and Porosity* (Third Edition). 1991 ISBN 0-412-39690-4
3. K. Heiskanen: *Particle Classification*. 1993 ISBN 0-412-49300-4
4. O. Molerus: *Principles of Flow in Disperse Systems*. 1993 ISBN 0-412-40630-6
5. I. Claus Bernhardt: *Particle Size Analysis*. Classification and Sedimentation Methods. 1994 ISBN 0-412-55880-7
6. Yasuo Arai: *Chemistry of Powder Production*. 1996 ISBN 0-412-39540-1
7. Terence Allen: *Particle Size Measurement* (Fifth Edition). 1996
   Volume 1: Surface area and pore size determination. ISBN 0-412-72950-4
   Volume 2: Powder sampling and particle size measurement. ISBN 0-412-75330-8
   Set of 2 vols., ISBN 0-412-75350-2
8. George E. Klinzing, Roy D. Marcus, Farid Rizk and L.S. Leung†: *Pneumatic Conveying of Solids*. A Theoretical and Practical Approach (Second Edition). 1997 ISBN 0-412-72440-5
9. Jonathan P.K. Seville, Uğur Tüzün and Roland Clift: *Processing of Particulate Solids*. 1997 ISBN 0-7514-0376-8
10. B. Kaye: *Powder Mixing*. 1997 ISBN 0-412-40340-4
11. O. Molerus and K.-E. Wirth: *Heat Transfer in Fluidized Beds*. 1997 ISBN 0-412-60800-6
12. R. Weinekötter and H. Gericke: *Mixing of Solids*. 2000 ISBN 0-7923-6229-2

KLUWER ACADEMIC PUBLISHERS – DORDRECHT / BOSTON / LONDON